Automotive Threat Analysis and Risk Assessment in Practice

Rodrigo do Carmo · Alexander Schlensog

Automotive Threat Analysis and Risk Assessment in Practice

A practical guide to TARA following the ISO/SAE 21434 standard for automotive embedded and IT/OT systems

Springer

Rodrigo do Carmo
Division Industry
secunet Security Networks AG
Eschborn, Germany

Alexander Schlensog
Division Industry
secunet Security Networks AG
Essen, Germany

ISBN 978-3-662-69613-2 ISBN 978-3-662-69614-9 (eBook)
https://doi.org/10.1007/978-3-662-69614-9

This Springer imprint is published by the registered company Springer-Verlag GmbH, DE, part of Springer Nature.
The registered company address is: Heidelberger Platz 3, 14197 Berlin, Germany

Foreword

They say that theory is second best to practice and this book will, by demonstrating application, empower the reader to master the underlying theory—thus proving the paradigm.

This book does not come out of nowhere. When I first tasked Dr. do Carmo to create the Component TARA (Threat Analysis & Risk Assessment) for an electronic brake system I only had a vague notion how deep the rabbit hole was that I was sending him down. ISO 21434 was not released at that time so he had to work his way with the existing body of (draft-) international standards and practices.

Through my own work and experiences in international standardization projects (ISO/SAE PAS 8475, ISO TR 8477 and ISO/IEC 5888) I bear witness that the meticulous design of the scopes of standards often challenge realization of projects. In business context project targets often create blurred lines between technical and standardization domains. In Dr. do Carmo's case those were Product Cybersecurity, Enterprise IT, and Operational Technology (OT). Capitalizing on his scientific research background he went to work and brought system and order to a vast deployment and risk landscape of PKIs, production servers, manufacturing execution systems, logistics and shipping systems, warranty workflows—just to name a few elements.

As a result of this activity he not only learned how to apply the relevant standards across application domains but also delivered the groundwork for consolidated risk management which greatly helped the company to optimize existing and conceive new business processes.

Going through this book as a former supervisor, proof-reader, and friend of the author I reminisced about past challenges and successes and developed fond anticipation that through reading and application of the presented knowledge others might be spared from painful and costly iterations.

Eschborn
27.10.2023

Michael Gerhard Schneider
Principal Expert Product Cybersecurity & Privacy,
Group Sector Automotive, Continental AG

The secret to successful and secure action lies in navigating the space between Hans-Olaf Henkel's assertion, 'If you want to eliminate all risk, you will destroy all opportunities', and the time-honored wisdom of 'Better safe than sorry'. This delicate balance involves managing risk, a complex task that necessitates weighing security, costs, benefits, and risk itself. Effective risk management often requires adherence to national and European (cybersecurity) legislation and standards, employing specialized risk analysis procedures and processes, while also accommodating the unique needs of individual companies.

Alexander Schlensog's expertise is precisely what makes him invaluable in this context.

Meeting Alexander nearly 15 years ago, I was tasked with accrediting a highly regulated, secure, and binding communication service. This new process demanded three certifications, demonstrating a high level of information security and data protection, as well as functionality and interoperability—a realm I was unfamiliar with.

Alexander's deep and comprehensive expertise in information security and risk management was instrumental. He excels at imparting his vast knowledge to varied audiences, ensuring the information is accessible at the right level. His pragmatic approach to navigating the challenges introduced by new legislation and regulation highlights his significant strengths. Alexander's ongoing responsibility for the security of the services I oversee and his advisory role in our numerous certifications is no mere coincidence.

His profound knowledge and ability to communicate effectively make Alexander an ideal candidate to author a book on risk management in highly regulated markets, such as the automotive industry. I eagerly anticipate the insights this book will offer and am convinced it will be indispensable for anyone deeply engaged with high-level risk analysis and management, extending even beyond industry boundaries.

Karlsruhe
14.02.2024

Leslie Romeo
Head of the ID & Trust Services division of
1 & 1 De-Mail GmbH, part of the United Internet Group

Preface

With the recent emergence of cybersecurity regulations in the automotive industry, a formalized approach to risk management has become mandatory. We begin by classifying these regulations and providing an overview of the European legal landscape in cybersecurity. Next, we highlight the relevant cybersecurity standards, with a focus on road vehicle production, while also considering the broader manufacturing industry where applicable.

Given the surge in risk management projects in the automotive sector, we, as authors, have gained significant experience by actively participating in numerous Threat Analysis and Risk Assessment (TARA) projects recently. Since the ISO/SAE 21434 has been released for the automotive industry, there is a well-defined approach to performing TARA. However, we have found that translating the standard's guidance into real-world projects is not always straightforward. During our extensive experience with TARA projects, we have repeatedly encountered fundamental technical discussions about certain aspects of TARA and the TARA process itself. This is particularly true for cybersecurity professionals, who lack expertise in risk management and threat modeling.

In this book, we aim to approach each step of the TARA process in a pragmatic manner. Throughout the book, we address three different domains of modern vehicles: classical TARA for individual electrical and electronic (E/E) systems within the vehicle, TARA at the vehicle level, including its connected ecosystem, and TARA for manufacturing plants. At the end of each chapter, which covers a fundamental step of TARA, we provide detailed case studies related to the above domains.

The first domain, the E/E systems within the vehicle and the vehicle itself, is the main focus of ISO/SAE 21434. However, ISO/SAE 21434 also requires that the vehicle and its components be protected during and after production. This requirement complements UN Regulation No. 155 on the vehicle level, which also enforces cybersecurity during and after production. To attain this comprehensive coverage, we adhere to the TARA methodology as outlined in the ISO/SAE 21434 standard, while incorporating certain aspects, such as the impact categories and the definition of items, assets, and threats, with established standards for information security (ISO/IEC 27000 series) and cybersecurity in the operational technology (IEC 62443). In no way do we attempt to harmonize or demonstrate interoperability between different standards in this book. Instead, our goal is to show how the generic TARA approach of ISO/SAE 21434 can be effectively applied to manufacturing systems, complementing the process with insights drawn from other sources.

Readers can expect a technical guideline that outlines how to proceed at each step of the TARA process. Our mantra is to perform TARA with technical rigor, while making the process as efficient as possible. We offer recommendations for optimizing the TARA process, such as providing insight on defining assets, determining the involvement of team members, suggesting approaches to defining damage scenarios, and identifying practical impact categories beyond those provided by the ISO/SAE 21434 standard. For instance, an essential aspect of the process is the identification of threat scenarios. Here, readers can expect practical advice based on well-defined approaches to threat modeling in the cybersecurity world. We present recommendations on how to work with existing threat catalogs. We delve into topics such as defining attack paths and determining the most effective methods for conducting feasibility evaluations. Toward the end of the book, we show how using statistical data from risks can enhance the overall understanding of TARA results, culminating in the final step of the TARA process: the risk treatment decision.

This book is intended primarily for cybersecurity engineers, with or without automotive experience, who are beginning to work on TARA projects. However, it is also a valuable resource for those already involved in automotive TARA projects who are keen to learn from our experience in specific areas of the TARA process. In addition, automotive cybersecurity teams seeking to perform TARA for manufacturing plants will find practical insights and knowledge sharing from our extensive experience. While this book is designed to be accessible, it does assume that readers have a foundational understanding of general cybersecurity concepts and principles.

We hope that this book serves as a valuable tool for conducting TARA, providing a more in-depth understanding of risk management, and equipping automotive professionals with the necessary expertise to efficiently tackle TARA projects.

Rodrigo do Carmo
Alexander Schlensog

Acknowledgements

I extend my sincere thanks to Prof. Dr.-Ing. Sebastian Biedermann and Manuel Wurm, who provided invaluable technical support and ideas for this book. I want to extend a special thanks to my colleague and friend, Michael Gerhard Schneider, for proofreading, providing ideas, and contributing the foreword to this book. Our journey together through the automotive cybersecurity jungle from day one has made this book possible.

I want to express a particular thank you to my co-author, Alexander Schlensog, whose extensive experience and profound knowledge in risk management and information security have significantly enriched the content.

Last but not least, writing a book is an extremely time-consuming endeavor alongside one's regular job. I would like to thank my wife Katharina and my son Maximo for their patience and support along the way.

Rodrigo do Carmo

My heartfelt thanks go to my longtime companions and friends, Mete Boz, Cord Kielhorn, and Werner Koch, who sparked my interest in information security and inspired me to pursue a path in this field. I would also like to extend my gratitude to Leslie Romeo and Roland Brüggemann for their intensive collaboration on particularly challenging projects. It is the exploration of the new and unknown that has always challenged me the most and spurred my development.

I am grateful to my many colleagues who have supported me tirelessly and kept me on the right path. Special thanks go to my esteemed colleague and co-author, Dr. Rodrigo do Carmo, who one day approached me with the idea for this book. Thank you for our collaboration, and I look forward to our future journey. I would also like to thank Justine Matzke and Dr. Hans-Jürgen Heinrich for their legal advice and proofreading.

Alexander Schlensog

Contents

Contents

About the Authors

Dr.-Ing. Rodrigo do Carmo received his degree in Telecommunications Engineering from Universidad Blas Pascal, Argentina, in 2010, and his doctoral degree in Computer Science from Technische Universität Darmstadt, Germany, in 2014. From 2015 he worked as Cyber Security Expert and Cyber Security Project Manager at Continental Teves AG & Co. oHG. Since 2020 he works at secunet Security Networks AG as Principal Consultant and Team Lead.

At Continental, Dr. do Carmo was responsible for the overall cybersecurity R&D activities for brake systems. Years before the ISO/SAE 21434:2021 standard appeared, Dr. do Carmo worked on cybersecurity processes, cybersecurity management and technical developments, including risk analysis and threat assessment. Dr. do Carmo also contributed to the review activities of ISO/SAE 21434:2021. He also coordinated the cybersecurity activities across the complete vehicle life cycle (key management) with a strong focus on manufacturing. He co-created a Center of Competence for cybersecurity for braking systems and he worked on several projects for German and international OEMs.

Since 2020, Dr. do Carmo provides consulting services to customers in the manufacturing industry, including automotive, ranging from small and medium suppliers to large manufacturers. He has led numerous TARA projects for the automotive and manufacturing industries, including TARA projects for IEC 62443. Dr. do Carmo is an active member of the ISO and DIN automotive committees working on cybersecurity and the VDMA committee for industrial security.

Dipl.-Phys. Alexander Schlensog studied physics at Gerhard-Mercator University in Duisburg and Heinrich-Heine University in Düsseldorf. He has worked in information technology since 2001 and began implementing and operating management systems in 2006.

Since joining secunet Security Networks AG in 2008, he has focused on information security consulting. As a Principal Consultant, he significantly influenced the consulting practice and took on responsibility for the development of the consulting business, notably contributing to the expansion in the area of critical infrastructures.

Since 2020, he has overseen the consulting business for the Industry division as its global head. As an expert in information security across information and operational technology, he specializes in highly regulated markets. Alexander Schlensog also serves on the ISO and DIN standards committees for information security and data protection.

Abbreviations

AC	attack complexity
ADAS	advanced driver assistance systems
AI	artificial intelligence
APT	advanced persistent threat
ASIL	automotive safety integrity levels
ASRG	Automotive Security Research Group
ATM	Auto-ISAC Automotive Threat Matrix
Auto-ISAC	Automotive Information Sharing and Analysis Center
AV	attack vector
BCP	business continuity planning
BIA	business impact analysis
BOM	bill of materials
BSI	Bundesamt für Sicherheit in der Informationstechnik[1]
CAL	cybersecurity assurance level
CAN	controller area network
CCC	Chaos Computer Club
CIA	confidentiality, integrity, availability
CLI	command-line interface
CNC	computer numerical control
CRA	EU Cyber Resilience Act
CSF	NIST Cybersecurity Framework
CSIA	cybersecurity interface agreement
CSR	certificate signing request
CSMS	cybersecurity management system
CVE	common vulnerabilities and exposures
CVSS	common vulnerability scoring system
CWE	common weakness enumeration
DA	delegated act
DDoS	distributed denial of service

[1] German Federal Office for Information Security.

DID	data identifier
DIS	draft international standard
DoIP	diagnostic over internet protocol
DoS	denial of service
DSSAD	data storage system for automated driving
DTC	diagnostic trouble code
E/E	electrical and electronic
ECU	electronic control unit
EDR	event data recorder
ENISA	European Union Agency for Cybersecurity[2]
ENX	European Network Exchange Association
EOL	end of line
ETL	ENISA Threat Landscape
ETSI	European Telecommunications Standards Institute
EU	European Union
HLS	high-level structure
HTTP	hypertext transfer protocol
FTP	file transfer protocol
FW	firmware
GDPR	EU General Data Protection Regulation
GNSS	global navigation satellite system
GW	gateway
HARA	hazard analysis and risk assessment
HMI	human-machine interface
HSM	hardware security moduly
HW	hardware
IACS	industrial automation control system
ICS	industrial control system[3]
ID	context-dependent, used both for *identifier* and *identification*
IDS	intrusion detection system
IEC	International Electrotechnical Commission
IMS	integrated management system
I/O	input/output
IP	internet protocol
IPS	intrusion prevention system
ISA	International Society of Automation
ISMS	information security management system
ISO	International Organization for Standardization
ISO-TP	ISO 15765-2 transport protocol
IT	information technology

[2] Formerly known as *European Network and Information Security Agency*.

[3] Used in this book as a synonym for IACS.

JTAG	Joint Test Action Group
LIN	local interconnect network
MAC	message authentication code
MITM	man-in-the-middle (attack)
MOST	media oriented systems transport
NIST	US National Institute of Standards and Technology
OBD	on-board diagnostic
OEM	original equipment manufacturer
OT	operational technology
OTA	over the air
PCB	printed circuit board
PII	personally identifiable information[4]
PKI	public key infrastructure
PLC	programmable logic controller
PM	permission
PR	privileges required
RAM	random-access memory
R&D	research and development
RDP	remote desktop protocol
RED	EU Radio Equipment Directive
RF	radio frequency
RQ	requirement
SAE	Society of Automotive Engineers
SBOM	software bill of materials
SCADA	supervisory control and data acquisition
SDV	software defined vehicle
SecOC	secure onboard communication
SHE	secure hardware extension
SMB	server message block
SoC	system-on-chip[5]
SOME/IP	scalable service-oriented middleware over internet protocol
SOTIF	safety of the intended functionality
SQL	Structured Query Language
SID	service identifier
SSH	secure shell protocol
SUC	system under consideration
SUMS	software update management system
SW	software
TAF	targeted attack feasibility
TARA	threat analysis and risk assessment

[4] Used in this book as a synonym for *personal data*.

[5] The acronym SOC is also used for *security operations center* in the literature.

TCU	telematic control unit
ToE	target of evaluation
TPMS	tire-pressure monitoring system
TVRA	threat vulnerability and risk analysis
UDS	unified diagnostic services
UI	user interaction
UN	United Nations
UNECE	United Nations Economic Commission for Europe
VCS	vehicle cybersecurity
VDA-ISA	VDA Information Security Assessment
V2V	vehicle-to-vehicle
V2X	vehicle-to-everything
VIN	vehicle identification number
WP	work product
XCP	universal measurement and calibration protocol

Chapter 1
Introduction

Abstract This introductory chapter lays the foundation for our book by providing an overview of the current state of regulation in the field of cybersecurity for road vehicles. We focus on presenting the key legal frameworks that are significant for manufacturers, including the EU Directive NIS-2, the UN Regulations No. 155 and No. 156, and other legal norms. These laws and regulations mark a significant step towards a more comprehensive consideration of cybersecurity in the automotive industry, which is taking a leading role in adapting to these new regulatory challenges. Furthermore, we provide a brief overview of the UN Regulation No. 155 and its relationship to the international standard ISO/SAE 21434.

Since its inception, the automotive industry has established itself as one of the most dynamic and adaptive sectors of the modern economy. Its history is one of constant innovation and transformation. The sector not only reacts to new developments, but also regularly acts as a catalyst for technological advances in the industry.

Road vehicles have advanced from simple mechanical constructs to highly complex and highly networked systems on wheels. Today, a single vehicle can have over a hundred control units that communicate with each other via special bus systems and protocols. This complexity itself represents a risk that needs to be managed.

This evolutionary step has opened up a new dimension of connectivity, in which vehicles are not only networked internally, but also externally. They interact with vehicle manufacturers' back-end systems, offer personalized services to occupants, communicate with other vehicles (V2V), and integrate into the wider infrastructure through vehicle-to-everything (V2X) communication.

The introduction of these cutting-edge technologies brings new challenges. New possibilities of attack, the complexity of individual components, and networking to form highly complex systems require the adaptation of existing procedures and often the development of new procedures and standards to ensure that the necessary security and safety can continue to be guaranteed.

It is by no means the case that the goal of ensuring security and safety is new, but the risk landscape has changed significantly. It will continue to change, and at a speed that has not been usual or customary in the automotive sector in recent decades. As a result, the requirements for the life cycle are also changing fundamentally. The

R. do Carmo and A. Schlensog, *Automotive Threat Analysis and Risk Assessment in Practice*, https://doi.org/10.1007/978-3-662-69614-9_1

consideration of (cyber) security has become indispensable to be able to achieve the goal of safety appropriately at all.

Legislators have also recognized that adapted regulations are necessary and have taken action. In the following chapters, we provide an overview of the current status of regulation and standardization.

1.1 European Regulation

In the following sections, we provide an overview of the current status of regulation and standardization.

1.1.1 Overview of Current Regulation and its Objectives

Regulation in the field of road vehicles has a long history. The *1958 Agreement* [1], more precisely the "Agreement Concerning the Adoption of Harmonized Technical United Nations Regulations for Wheeled Vehicles, Equipment and Parts which can be Fitted and/or be Used on Wheeled Vehicles and the Conditions for Reciprocal Recognition of Approvals Granted on the Basis of these United Nations Regulations", was drawn up under the auspices of the UNECE at the proposal of the Federal Republic of Germany and came into force in 1959. The aim of the agreement was to create uniform procedures for the approval of motor vehicles. Technical requirements and test procedures were defined, which must be fulfilled by manufacturers and suppliers to achieve (type) approval. Of particular importance is the goal of "safety", i.e., ensuring the protection of occupants and all other road users. With the rapidly increasing networking of motor vehicles, the use of digital elements (e.g., control units), and new technologies, new classes of risk are emerging. With a delay of a few years, regulation is now reacting visibly and clearly, and not just in the automotive sector. The manufacturing industry as a whole is being forced to integrate risks relating to cybersecurity and the impact on safety into its thinking and actions.

It is not the case that safety in the sense of security or safety has been ignored in the past. However, the risk landscapes and related requirements are increasingly shifting towards "(cyber) security for safety". Cybersecurity must be considered, evaluated, and dealt with in systems with digital elements to achieve the goal of safety.

In the following, we provide an overview of the key cybersecurity legislation projects that focus on manufacturing companies and manufactured products, e.g., for the approval of electronic components. We look at the manufacturing industry in general and the automotive industry in particular, as it plays a pioneering role here. There are other industries, such as medical device manufacturers or manufacturers in the nuclear energy sector, that are subject to individual legal standards similar to the automotive industry. These are not the focus here, but are nevertheless comparable in terms of the challenges, possible approaches, and special methods, such as the TARA discussed in this book.

1.1.2 UN Regulation No. 155

UN Regulation No. 155 [2] defines uniform provisions for the cybersecurity of road vehicles. The regulation stipulates that a cybersecurity policy must be implemented and a cybersecurity management system set up. The appropriateness and effectiveness of this system must be proven for the type approval of new vehicles. The system supports vehicle manufacturers and suppliers with processes and tools to systematically identify, assess, and manage cybersecurity risks. The ISO/SAE 21434 standard helps companies to implement parts of the requirements.

1.1.3 UN Regulation No. 156

UN Regulation No. 156 [3] stipulates a Software Update Management System (SUMS) to ensure that software updates in vehicles are carried out securely and in compliance with the law. In particular, the regulation is intended to prevent dangerous software malfunctions and manipulation during the update process. In the event of a failed update, the software must either continue to function properly or assume a safe state. Manufacturers are required to set up a system for updating vehicle software that is safe and reliable. The regulation also stipulates that manufacturers must provide the authorities with a detailed report on the software updates and the changes they make. The regulation is intended to ensure that software updates do not compromise the safety and environmental performance of the vehicle, i.e., do not have a negative impact on the environment.

1.1.4 EU NIS-2

The EU NIS-2 Directive (EU Directive on Security of Network and Information Systems) [4] is part of the EU Cybersecurity Strategy[1, 2] and is intended to create a high level of cybersecurity at the EU level and strengthen the internal market. It is a response to an ever-increasing threat situation in cyberspace and consolidates the ability to act at the EU level. The directive is the further development of the EU NIS Directive (Directive (EU) 2016/1148) adopted in 2016, the implementation of which is known in Germany as the *IT-Sicherheitsgesetz*.

[1] https://digital-strategy.ec.europa.eu/en/policies/cybersecurity-strategy

[2] https://digital-strategy.ec.europa.eu/en/library/eus-cybersecurity-strategy-digital-decade-0

The primary objective is to improve the level of cybersecurity and create a high common level to strengthen the EU internal market. The powers of the competent authorities will be extended. Action should be effective, proportionate, and dissuasive. Affected companies are obliged to implement active risk management and comply with the state of the art, not only for themselves but also regarding their supply chains—a late learning that came out of various crises. The "digital" world should be considered and secured in conjunction with the "physical" world. Management bodies will be held liable, including personally.

The new directive has changed the scope of application so that the focus is no longer on critical infrastructures but encompasses large parts of the economy. The manufacturing industry and manufacturers of goods are classified as important. However, the mobility sector itself is also part of the energy transition and, at the same time, a building block for smart cities. This networking with the municipal infrastructure therefore has a direct and indirect impact on the transportation and energy sectors. Cybersecurity risks and measures must be seen in the overall context to avoid any gaps in consideration.

The ISO/IEC 27001 standard offers the best basis for implementing a management system due to its high level of international acceptance. The standard makes it possible to implement[3] the requirements of the directive with regard to cybersecurity and a comprehensive active risk management system.

1.1.5 EU Cyber Resilience Act (CRA)

The EU Cyber Resilience Act (Regulation on horizontal cybersecurity requirements for products with digital elements) forms the legal framework that describes the cybersecurity requirements for hardware and software products that are placed on the market in the European Union. It obliges manufacturers to ensure cybersecurity throughout the life cycle of a product, from design to operation.

Until now, various legal acts and initiatives at national and union levels have only partially addressed the challenges of cybersecurity and have done so in different ways, leading to unnecessarily high costs. The new regulation creates a coherent cybersecurity framework for hardware and software manufacturers.

The regulation is another element of the EU Cybersecurity Strategy. The main objective of the regulation is to strengthen cybersecurity and the smooth functioning of the internal market. It creates the conditions for the development of secure products with digital elements, and thus prevents cybercrime.

---

[3] By implementing the directive, it is meant that the requirements of national implementing laws can be fulfilled through ISO/IEC 27001. In Germany, for example, this refers to the "NIS2-Umsetzungs- und Cybersicherheitsstarkungsgesetz".

The regulation addresses two main problems: firstly, the current often low level of cybersecurity of products with digital elements, i.e., hardware and software, is reflected in widespread vulnerabilities and the inadequate provision of security updates. Secondly, the inadequate understanding and access of users to information on the products addressed. This prevents buyers from selecting products with suitable cybersecurity features and operating them securely.

Products with digital elements that are integrated into or can be connected to information systems are a potential attack vector for malicious actors such as hackers or nation-states. Viewed on their own, as non-critical, such products enable information systems or networks to be compromised under certain circumstances.

For this reason, security requirements must be taken into account right from the design and development phase. In other words, products should be "secure by design and default". Companies must create secure products in a risk-oriented manner and ensure effective vulnerability management combined with active communication. In addition, comprehensive documentation on the secure use of products must be provided. New security threats must be dealt with appropriately, and any gaps that have arisen must be closed quickly with software patches.

The regulation is aimed at all companies that bring products with digital elements into the European Economic Area, but certain areas are excluded. These include the military sector, civil aviation, and air traffic control, but also the automotive industry, as the type approval process provides sufficient protection here.

The IEC 62443 series of standards can generally be used to implement the regulation, which helps to establish an appropriate cybersecurity management system within organizations and along the supply chain. Moreover, it is also suitable for certifying systems or individual components and thus proving cybersecurity over the life cycle. The ISO/SAE 21434 standard is intended for the specific area of road vehicles, which pursues precisely these objectives with the difference that it is not individual elements or systems that are certified here, but organizations. There are other similar "replacement" standards for other sectors of the economy.

Related to this is the Radio Equipment Directive (RED), which is intended to ensure "a high level of protection in the fields of health and safety, an adequate level of electromagnetic compatibility, and an effective and efficient use of the radio spectrum to avoid harmful interference" in Europe. The RED was supplemented by the Delegated Act (DA) 2022/30, which aims to protect so-called "internet-enabled radio devices"[4], (mobile) networks and personal data. The increasing risk situation in the area of cybersecurity and data protection is also a key consideration here.

[4] Internet-enabled radio devices are those devices that can communicate via the Internet themselves, regardless of whether they communicate directly or via other devices ("Internet-connected radio devices").

1.1.6 EU General Safety Regulation

The EU General Safety Regulation[5] [5] is a European regulation with the aim of improving road safety and reducing the number of road accidents. It lays down minimum requirements for the safety of road vehicles and trailers, as well as systems, components, and separate technical units for these vehicles and the protection of vehicle occupants and other vulnerable road users.

Regarding cybersecurity, the regulation addresses the protection and exchange of safety-related data between vehicles and infrastructures. This involves the collection, storage, and exchange of data that is intended to both protect the privacy of users and improve road safety. The regulation lays the foundation for robust cybersecurity practices and measures to ensure the integrity and security of vehicle systems and data, including in relation to autonomous driving systems and driver assistance systems.

The regulation also emphasizes the importance of road infrastructure safety and calls for measures to improve road safety. At this point, it should be pointed out once again that road infrastructure is one of the critical infrastructures that must be protected in a special way. This regulation is of great importance for engineers and manufacturers in the automotive sector, as it promotes the development of safe vehicles and systems and, at the same time, supports the introduction of innovative technologies, particularly in the area of information security and cybersecurity.

The ISO 26262 standard is an important element for the implementation of safety requirements in the automotive sector. There are a number of other safety standards for other sectors, e.g., IEC 60601-1 for medical electronic devices or IEC 61511 for the process industry.

1.1.7 EU Machinery Regulation

The aim of the EU Machinery Regulation[6] is to bring the safety requirements for machinery up-to-date and harmonize them. It creates a new legal framework for placing safe machinery and products on the EU market. New risks associated with current and emerging technologies, particularly digital technologies such as the Internet of Things or artificial intelligence, are considered. The regulation defines the essential health and safety requirements that machinery products must meet to be placed on the European market and sets out the procedures for assessing their conformity. The new regulation requires protection against cyberattacks. Machines and products must be designed and constructed in such a way that cyberattacks do not pose a risk to life, limb or the environment. In addition, the term "safety components" has been

---

[5] The regulation is a legal act relating to the EU type-approval procedure.

[6] Regulation (EU) 2023/1230 on machinery (which replaces the "Directive 2006/42/EC on machinery" as of 20.01.2027).

redefined and now also includes software that serves this purpose and is placed on the market separately.

1.1.8 EU General Data Protection Regulation

The EU General Data Protection Regulation (GDPR) [6] is a legal framework of the European Union that came into force on May 25, 2018, and regulates the protection of personal data of EU citizens. The regulation ensures that companies and organizations that process personal data have clear accountability and transparency regarding the use and protection of that data. The GDPR stipulates that data processors and data controllers must take appropriate technical and organizational measures to ensure an appropriate level of data protection.

Compliance with technical and organizational requirements is of central importance here. To meet these requirements, companies, and organizations in the EU must, among other things, implement appropriate and effective cybersecurity strategies and systems. This includes, for example, the encryption of data, regular security checks and the training of staff in the secure handling of personal data. In addition, data breaches must be reported to the relevant supervisory authorities.

The requirements of the regulation make cybersecurity an integral part of data processing practices within the EU. To implement the regulations, it is necessary to set up and operate a data protection management system. The ISO/IEC 27701 standard can be used for this purpose.

1.2 UN Regulation No. 155 and International Standard ISO/SAE 21434

The UN Regulation No. 155 [2] is a comprehensive regulatory framework developed by the United Nations Economic Commission for Europe (UNECE) to establish a unified approach to cybersecurity in vehicles. The regulation is an extension of the *1958 Agreement*, which defines minimum technical requirements for type approval and homologation of vehicles. Although the agreement was originally developed for Europe (specifically ECE members), the agreement now includes the participation of many non-ECE members. While UN Regulation No. 155 primarily applies to passenger vehicles, specifically those falling under categories *M* (vehicles with at least four wheels), *N* (vehicles for transporting goods), and *O* (trailers with at least one ECU), it also extends to categories *L6* and *L7* if the vehicles possess level 3 or higher automated driving functions. The countries involved in the 1958 Agreement are responsible for incorporating the regulation into their national laws.

The UN Regulation No. 155 has essential importance for promoting cybersecurity in the automotive industry, because it ensures the cybersecurity of vehicles throughout their complete life cycle, including development, production, operation, and maintenance. The key requirement of the regulation is that car manufacturers are required to implement and maintain a *Cybersecurity Management System (CSMS)*. Essentially, a CSMS requires carmakers to implement and verify processes in the areas of cybersecurity governance, cybersecurity culture, risk management, validation and verification, data monitoring for forensic analysis, and supplier management. Carmakers must ensure that cybersecurity is addressed throughout the supply chain. Specifically for risk management, risks in the vehicle must be identified and assessed, and protective measures must be implemented to address those risks.

The international standard ISO/SAE 21434 [7], developed by the International Organization for Standardization (ISO) in conjunction with the Society of Automotive Engineers (SAE), provides process requirements for achieving cybersecurity *risk management* across the complete vehicle life cycle, including the concept, product development, production, operation, maintenance, and decommissioning of E/E systems in road vehicles. ISO/SAE 21434 covers various aspects of cybersecurity, including organizational cybersecurity management, project dependent cybersecurity management, risk assessment, validation and verification techniques, supplier management, and continual cybersecurity activities. ISO/SAE 21434 provides a common framework for communicating and managing cybersecurity risks. The relationship between the UN Regulation No. 155 and the ISO/SAE 21434 across the supply chain is depicted in Fig. 1.1. Car manufacturers and their suppliers can achieve compliance with the risk management requirements of the UN Regulation No. 155 by implementing ISO/SAE 21434 and aligning the activities through cybersecurity interface agreements.

Focus on ISO/SAE 21434
This book focuses primarily on the threat analysis and risk assessment methodology of ISO/SAE 21434. While UN Regulation No. 155 is referenced throughout the book, particularly when addressing the threats specified in the regulation and conducting risk assessments during the vehicle production phase, it is important to note that this book can be used independently, solely for compliance with ISO/SAE 21434, regardless of any specific regulatory context.

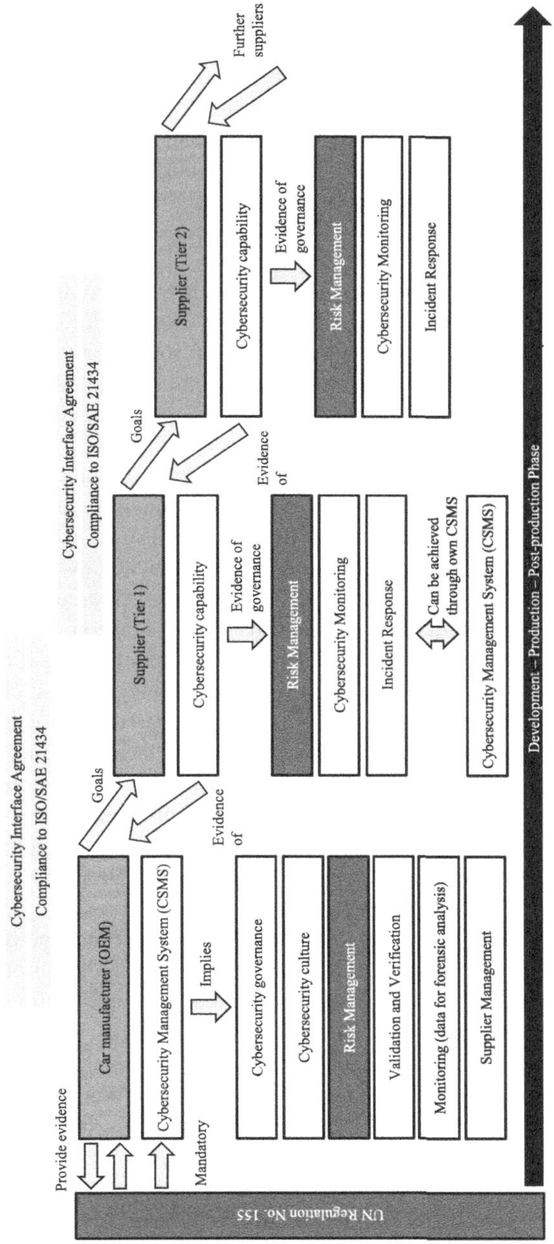

Fig. 1.1 Relationship between the UN Regulation No. 155 and the ISO/SAE 21434 across the supply chain

1.3 Structure of This Book

In Chap. 1, we lay the foundation for our book by providing an overview of the current state of regulation in the field of cybersecurity for motor vehicles. We focus on presenting the key legal frameworks that are significant for manufacturers, including the EU Directive NIS-2, the UN Regulations No. 155 and No. 156, and other legal norms. These laws and regulations mark a significant step towards a more comprehensive consideration of cybersecurity in the automotive industry, which is taking a leading role in adapting to these new regulatory challenges. Furthermore, we provide a brief overview of the UN Regulation No. 155 and its relationship to the international standard ISO/SAE 21434.

In Chap. 2, we provide an overview of standards for the manufacturing industry and specifically the automotive industry that are relevant for the compliance and implementation of the previously presented legal norms. We look at established and newly developed standards that can and, in some cases, must be applied not only in the automotive industry but also across the entire manufacturing sector, including ISO/SAE 21434, ISO/IEC 27001, IEC 62443, and TISAX®. These standards are significant in the context of the increasing networking and automation in vehicles and the general trend towards smart factories and products, to achieve cybersecurity.

In Chap. 3, we examine the characteristics of various standards in cyber and information security and (cyber) risk management. We provide an overview of the risk management methods and systems of the previously considered standards, with the aim of assessing their fundamental suitability for integration into a unified (risk) management system. We do not delve into technical details or legal restrictions. We consider practical aspects and possible limitations in integrating such systems.

In Chap. 4, we provide a more in-depth introduction to the threat analysis and risk assessment process. We highlight the significance of TARA within the automotive industry, explaining its purpose and necessity. We also describe when a TARA needs to be performed in the vehicle life cycle, based on our professional experience and the regulatory framework. We also discuss the duration and level of technical detail of a TARA project according to the life cycle phase. The overall TARA process covered in this book is described in the following chapters.

In Chap. 5, we define the scope of a TARA project. The scope defines how the TARA will be addressed. In addition to describing the item definition and its operational environment, at this step we describe technical assumptions for the TARA, e.g., about existing controls, existing vulnerabilities, cybersecurity goals, or cybersecurity claims. Any special features of the TARA project are described in the scope. In addition, we describe the attacker model and its capabilities. We also define the impact categories for the damage scenarios in the context of the scope. In this chapter we also introduce the practical case studies that will be presented throughout the book.

In Chap. 6, we lay the groundwork for a TARA by identifying the assets of the system under analysis, or item. We provide practical insights into asset identification and discuss relevant cybersecurity properties in addition to the CIA triad. In addition, we provide a valuable list of common assets for both automotive embedded and

IT/OT systems. Following asset identification, we move on to identifying damage scenarios. We analyze the damage scenarios affecting the road user and provide practical examples of how to evaluate the impact across the categories required by the standard. Following the ISO/SAE 21434 standard, we show how we can extend the view of the damage scenario to analyze the impact on an organization as a whole. For the business impact, we describe new impact categories, such as intellectual property and image/reputation. At the end of the chapter, we provide a complete asset identification for three practical case studies.

In Chap. 7, we tackle the identification of threat scenarios. Identifying threat scenarios and assigning them a probability of occurrence (feasibility) is the quintessence of a TARA. In this chapter, we address several key challenges in performing TARA. We describe how to actually perform threat identification and discuss the various popular threat modeling approaches. We also discuss which threat catalogs to use and how to use them. We also look at the attack paths and discuss the technical depth required and how to perform the attack feasibility evaluation. We conclude the chapter with our three real-world case studies.

In Chap. 8, we address the calculation of risk. Risk is derived from the combination of the feasibility rating of the threat scenarios and the impact of the threat scenario actually occurring. We perform the mapping of asset-threat scenarios to damage scenarios and calculate the risk of each combination. We provide practical insights on how to adjust the risk matrix. We also show how statistical representations can be useful to better present and understand the key results of a TARA. Finally, we provide examples based on our three case studies.

In Chap. 9, we finalize the TARA process with the risk treatment decision. Once the risk assessment has been completed, the next step is to decide what actions to take to address the resulting risks. In this step, the cybersecurity team decides, in accordance with ISO/SAE 21434 and corporate policy, whether the risks can be avoided, reduced, shared, or accepted. Each individual risk that is greater than a minimum value must be analyzed. If cybersecurity controls are applied to reduce the risks, the new risk values are calculated. In this step of TARA, cybersecurity claims and cybersecurity goals are defined when risks are accepted or shared, or when cybersecurity controls are applied. We continue with our case studies and provide examples of various risk treatment decisions, including the mapping of cybersecurity controls. Although this is the final formal step of the TARA, we describe how the TARA is actually an evolving document, and the process can be iterated as often as needed.

1.4 Naming Convention

In this book, we use the terms from ISO/SAE 21434:2021 [7] as they are. This includes terms such as *asset*, *attack feasibility*, *attack path*, *component*, *cybersecurity claim*, *cybersecurity goal*, *cybersecurity property*, *damage scenario*, *item*, *operational environment*, *threat scenario*, *weakness*, among others. We recommend consulting [7] for the exact definition.

Specific terms from other standards are properly cited when used. Generic terms such as *security*, *information security*, *IT security*, and *cybersecurity* are used indistinguishably, except when they belong to a term from ISO/SAE 21434, or when specific term accuracy is required.

References

1. United Nations: Agreement Concerning the Adoption of Harmonized Technical United Nations Regulations for Wheeled Vehicles, Equipment and Parts which can be Fitted and/or be Used on Wheeled Vehicles and the Conditions for Reciprocal Recognition of Approvals Granted on the Basis of these United Nations Regulations E/ECE/TRANS/WP.29/2016/2 (2017). https://unece.org/trans/main/wp29/wp29regs
2. United Nations: UN Regulation No. 155—Cyber security and cyber security management system E/ECE/TRANS/505/Rev.3/Add.154 (2021). https://unece.org/transport/documents/2021/03/standards/un-regulation-no-155-cyber-security-and-cyber-security
3. United Nations: UN regulation no. 156—software update and software update management system E/ECE/TRANS/505/Rev.3/Add.155 (2021). https://unece.org/transport/documents/2021/03/standards/un-regulation-no-156-software-update-and-software-update
4. European Union: Directive (EU) 2022/2555 of the European Parliament and of the Council of 14 December 2022 on measures for a high common level of cybersecurity across the Union, amending Regulation (EU) No 910/2014 and Directive (EU) 2018/1972, and repealing Directive (EU) 2016/1148 (NIS 2 Directive) PE/32/2022/REV/2 (2022). https://eur-lex.europa.eu/eli/dir/2022/2555
5. European Union: Regulation (EU) 2019/2144 of the European Parliament and of the Council of 27 November 2019 on type-approval requirements for motor vehicles and their trailers, and systems, components and separate technical units intended for such vehicles, as regards their general safety and the protection of vehicle occupants and vulnerable road users, amending Regulation (EU) 2018/858 of the European Parliament and of the Council and repealing Regulations (EC) No 78/2009, (EC) No 79/2009 and (EC) No 661/2009 of the European Parliament and of the Council and Commission Regulations (EC) No 631/2009, (EU) No 406/2010, (EU) No 672/2010, (EU) No 1003/2010, (EU) No 1005/2010, (EU) No 1008/2010, (EU) No 1009/2010, (EU) No 19/2011, (EU) No 109/2011, (EU) No 458/2011, (EU) No 65/2012, (EU) No 130/2012, (EU) No 347/2012, (EU) No 351/2012, (EU) No 1230/2012 and (EU) 2015/166 PE/82/2019/REV/1 (2019). https://eur-lex.europa.eu/eli/reg/2019/2144/oj
6. Regulation (EU) 2016/679 of the european parliament and of the council of 27 april 2016 on the protection of natural persons with regard to the processing of personal data and on the free movement of such data, and repealing directive 95/46/EC (general data protection regulation) (2016). https://eur-lex.europa.eu/eli/reg/2016/679/oj
7. ISO/SAE 21434:2021 Road vehicles—Cybersecurity engineering. Standard, International Organization for Standardization/Society of Automotive Engineers, Geneva, CH (2021). https://www.iso.org/standard/70918.html

Chapter 2
Standards for the Manufacturing and Automotive Industries

Abstract In this chapter, we provide an overview of standards for the manufacturing industry and specifically the automotive industry that are relevant for the compliance and implementation of the previously presented legal norms. We look at established and newly developed standards that can and, in some cases, must be applied not only in the automotive industry but also across the entire manufacturing sector, including ISO/SAE 21434, ISO/IEC 27001, IEC 62443, and TISAX®. These standards are significant in the context of the increasing networking and automation in vehicles and the general trend towards smart factories and products, to achieve cybersecurity.

The legal standards listed in the first chapter have many similarities; they pursue comparable and overlapping objectives and impose comparable requirements. To achieve the greatest synergy effects, the legal standards applicable to an organization should be considered and implemented in context.

For the automotive industry, continuous further development with comparatively regular development cycles and innovations is part of everyday life. Over the past two decades, we have witnessed a consistent and recently rapid rise in networking and automation in vehicles. This has prompted the creation of standards for harmonization and regulation in dealing with aspects such as vehicle safety or cybersecurity. This also applies to the entire manufacturing industry, described in Germany by the term "Industry 4.0". One could generally speak of "smart" factories and products.

In this chapter, we show which standards can, and in some cases must, be used in industry to implement the legal norms. We provide a brief overview of established and newly developed standards for the manufacturing industry, with a focus on the automotive industry.

R. do Carmo and A. Schlensog, *Automotive Threat Analysis and Risk Assessment in Practice*, https://doi.org/10.1007/978-3-662-69614-9_2

2.1 ISO/SAE 21434

ISO/SAE 21434 [1] is an international standard that was developed specifically regarding the cybersecurity of road vehicles and considers the increasing digitalization and networking of vehicles and the associated cybersecurity risks[1]. The standard standardizes language, objectives, requirements, and regulations in relation to cybersecurity and is intended to help with international harmonization.

The standard provides a foundation for the development of secure vehicles and connected systems, through a structured approach to identifying and mitigating cybersecurity risks across the entire life cycle and supply chain—from design, through development and production, to operation and decommissioning. By complying with this standard, automotive manufacturers and suppliers can systematically and sustainably establish, maintain, and improve cybersecurity in their products and services. The standard is related to UN Regulation No. 155 and supports organizations in implementing it to meet legal and regulatory requirements for type approval. The standard defines requirements for an organization-wide management system, but only the management of cybersecurity risks in road vehicles is mandatory. The application of the standards is therefore an important building block in the implementation of the UN Regulation No. 155, but it does not cover all requirements.

2.2 IEC 62443

IEC 62443 [2–10] is a series of standards developed specifically for the safety of industrial automation and control systems and networks (IACS). It arose from the realization that industrial systems, which were previously considered isolated, were and are becoming increasingly networked. This also makes this area more susceptible to cyberattacks. Originally developed by the International Society of Automation (ISA) under the name ISA99, it was later transformed into IEC 62443 and further developed and standardized by the International Electrotechnical Commission (IEC).

The series of standards aims to provide a structured approach to cybersecurity in industrial environments, often referred to as operational technology (OT). This includes the protection of equipment and systems from cyberattacks, which can endanger the safety of equipment and personnel as well as cause operational disruption and data loss. The series of standards prioritizes the security property of availability over integrity and confidentiality. It describes a CSMS[2] that includes risk analysis, the implementation of countermeasures, and monitoring and improvement. The series of standards is not only aimed at operators, but also at integrators, maintenance engineers of industrial machinery and suppliers or manufacturers.

[1] ISO/SAE 21434:2021 cancels and replaces the SAE J3061:2016 standard.

[2] The concept of CSMS was removed from IEC 62443 in the second edition of IEC 62443-2-1 (August 2024). The standard now requires coordinating the security program with the organization's ISMS or introducing a suitable process if no ISMS exists.

The series of standards is divided into four main categories, or "pillars", that deal with different aspects of industrial cybersecurity. These pillars are not to be understood as separate entities, but as part of an overarching framework designed to ensure security throughout the life cycle of an industrial automation system, from manufacture to integration to operation.

The first pillar (IEC 62443-1) deals with the general process of cybersecurity. Framework conditions, terminology, and concepts are defined. It also lays down the foundations for a security policy and the management of cybersecurity. This area is about establishing a culture of cybersecurity, which forms the basis for all further measures.

The second pillar (IEC 62443-2) focuses on policies and procedures, i.e., the processes with which cybersecurity is to be achieved and maintained. It includes guidelines for risk assessments, system classifications, and security levels. The pillar emphasizes the importance of taking a methodical approach.

The third pillar (IEC 62443-3) deals with the concept of systems. A system typically consists of a combination of hardware and software and is intended to perform control and automation tasks in industrial environments. This involves requirements that must be met during design, development, integration, and operation. Risk analyses are used to define measures, including the safety levels to be achieved. The safety levels are also defined by the IEC 62443 standard.

The fourth and final pillar (IEC 62443-4) deals with the requirements for the components of an industrial automation and control system. It is aimed in particular at manufacturers of such elements and helps to make the development processes and the safety functions of the components safe. Components are embedded devices, host devices, network components or software products. The aim is to ensure that the individual components used in an industrial system have the required security features and are designed in such a way that they offer effective protection against cyber threats.

Overall, the series of standards represents a holistic approach to ensuring appropriate and effective cybersecurity for industrial systems and networks throughout the supply chain and life cycle. Several standards from the family are certifiable, so that effective implementation can be demonstrated, from the organization to processes to systems or components.

2.3 ISO/IEC 27001

ISO/IEC 27001 [11] is an internationally recognized standard for the management of information security. The standard specifies the requirements for the establishment, implementation, maintenance and continuous improvement of an information security management system (ISMS). An ISMS is a systematic approach to protecting information by classifying it, identifying and evaluating risks that threaten these assets and implementing appropriate security objectives (controls).

The standard consists of two parts, the so-called standard body or standard framework and Annex A[3]. The body of the standard comprises chapters 4 to 10 and describes a process-oriented approach to the design, integration and management of an ISMS in the form of requirements. This includes aspects such as asset management, risk assessment, risk treatment, monitoring and improvement. Annex A contains a list of controls (translated as measures in the German standard) that can be implemented to reduce specific risks. A key strength of an ISMS in accordance with ISO/IEC 27001 is the enormous degree of freedom in its specific implementation. Essentially, objectives are predefined and the organization itself can determine the manner of implementation. No specific technologies or solutions are prescribed.

Organizations can be certified in accordance with the standard and thus provide proof that they have established and are continuing to develop an effective and appropriate management system.

2.4 ISO 26262

ISO 26262 [12] is an internationally recognized standard with the purpose of defining and ensuring the functional safety of electrical and electronic systems in motor vehicles.

A central element of the standard is the definition of safety objectives and the introduction of Automotive Safety Integrity Levels (ASIL). These ASILs categorize risks and are determined by a risk analysis of potential hazards based on three variables: severity, likelihood of exposure, and controllability by the driver. The standard considers the entire life cycle of safety-related systems, from initial conception to decommissioning. Specific requirements are placed on the organization to ensure that the functional safety objectives can be achieved. This includes the definition of roles, responsibilities, and necessary qualifications. During the product development process at system level, ISO 26262 specifies processes and activities ranging from safety analysis and the definition of safety objectives to the development of safety concepts. In addition, there are specific requirements for hardware and software development to ensure that these components meet the defined security objectives. Furthermore, the standard specifies requirements for the production and operational phases to ensure that functional safety objectives are maintained throughout the entire life cycle of a vehicle. Supporting processes such as configuration and change management, the qualification of software tools, and the qualification of hardware components are also part of the standard.

Overall, this standard provides a comprehensive framework that helps automotive manufacturers and suppliers to develop products that ensure an acceptable level of risk in relation to potential malfunction hazards.

[3] The ISO/IEC 27002 standard contains recommendations and instructions for implementing the controls from Annex A of ISO/IEC 27001.

2.5 ISO/SAE PAS 8475

The standard is being developed in order to deal with the security of data in automated and networked vehicle systems, and it is a development in collaboration between the International Organization for Standardization (ISO) and the Society of Automotive Engineers (SAE)[4]. The concept of the Cybersecurity Assurance Level (CAL) from ISO/SAE 21434 is included and extended by the idea of Targeted Attack Feasibility (TAF). CALs are assessment levels that can be used to indicate the robustness of vehicle systems or vehicles against cyberattacks. TAF is intended to help assess how realistic and feasible an attack on vehicle systems or vehicles is. This includes the complexity of the systems, the ability of attackers, and the effects of a successful attack. The TAF concept supports the planning and evaluation of the security of the systems under consideration. Both concepts can be used as part of the TARA process introduced by ISO/SAE 21434.

2.6 ISO PAS 5112

The ISO PAS 5112 [13] standard is a guideline for auditing the cybersecurity of road vehicles. It supplements ISO 19011 [14], which is a guide for auditing management systems and sets out general requirements. ISO PAS 5112 establishes criteria based on the ISO/SAE 21434 standard and thus enables an appropriate and effective audit of this standard.

The standard deals with the management of audit programs, the performance of audits and competence requirements for auditors and their review The standard supports the user in internal audits (1st party), supplier audits (2nd party) and certification audits (3rd party)[5].

2.7 ISO/SAE TR 8477

The standard is being developed as part of ISO/SAE 21434:2021 and deals with technical considerations for the planning and implementation of the verification and validation of the cybersecurity of components and systems in road vehicles[6].

---

[4] The standard is currently under development and its status can be viewed at https://www.iso.org/standard/83187.html.

[5] The German Association of the Automotive Industry (VDA) has published a questionnaire and an assessment scheme in the VDA volume Automotive Cybersecurity Management System Audit (ACSMS) December 2020. This can be used for auditing the CSMS (both for OEMs and contractual partners).

[6] The standard is currently under development and its status can be viewed at https://www.iso.org/standard/83188.html.

ISO/SAE TR 8477 is also closely linked to the concepts of Cybersecurity Assurance Levels (CAL) and Target Attack Feasibility (TAF), which are used in ISO/SAE PAS 8475.

2.8 ENX and TISAX®

ENX (European Network Exchange) is an organization founded by European automotive manufacturers and suppliers. ENX aims to standardize and secure communication and data exchange in the automotive industry. By providing a secure network, it aims to facilitate collaboration and ensure data integrity.

TISAX® (Trusted Information Security Assessment Exchange) is a standard for information security that was developed specifically for the automotive industry. TISAX® enables companies in the automotive industry to assess and validate compliance with security standards and requirements. The TISAX® assessments are carried out by accredited assessment service providers and are recognized among the participants of the TISAX® exchange. This facilitates cross-company collaboration and significantly reduces the effort required for individual security assessments.

Although the automotive industry is familiar with legislative requirements in many respects, specific regulations on information or cybersecurity have so far only been sporadic. For example, the German Association of the Automotive Industry (VDA) has developed a catalog for Information Security Assessments (VDA-ISA) in response to the OEMs' desire for more transparency regarding the information security level of suppliers. OEMs can then use the TISAX® model based on this catalog to check how suppliers handle sensitive data, for example, during the development of prototypes. In addition to the aspects of information security in TISAX®, ISO/SAE 21434 now sets requirements for the cybersecurity of the products themselves.

How do the regulations and standards influence TARA? The VDA published an updated version of its industry standard for information security assessments—the VDA Information Security Assessment (VDA-ISA) catalog version 6.0 [15].

To ensure industry-wide security standards, the VDA recommends that all companies involved in the automotive industry value chain develop their information security in accordance with the VDA-ISA.

The TISAX® standard is used for the cross-company assessment, recognition, and tracking of VDA Information Security Assessment results. TISAX® is therefore the gold standard for certifying information security in companies, particularly in the automotive industry. But even outside the automotive industry, meeting the requirements of TISAX® can help to protect important company assets in the form of data and information.

2.9 VDA Information Security Assessment

VDA Information Security Assessment (VDA-ISA) [15] is an integral part of the TISAX® standard. VDA-ISA refers to the assessment of a company's information security to ensure that it has implemented the necessary safeguards to protect sensitive information. As part of TISAX®, VDA-ISA is carried out according to a standardized procedure based on the VDA-ISA catalog, a catalog of security requirements and measures developed by the VDA. The implementation of the requirements is assessed because of maturity levels (from 0 to 5). The structure and content are based on ISO/IEC 27001, but do not cover it completely and are extended to include automotive industry-specific issues, such as prototypes and data protection. Suppliers can use VDA-ISA to prove their suitability in terms of cybersecurity.

2.10 ENX VCS

ENX has introduced a standardized audit program with the Vehicle Cybersecurity (VCS) Audit scheme [16]. The scheme is aimed at suppliers to the automotive industry who develop, produce or maintain E/E systems. A CSMS in accordance with ISO/SAE 21434 as defined by ISO PAS 5112 can be verified by means of auditing in accordance with ENX VCS. The certifications are therefore not automatically recognized internationally, but within the industry organized in ENX.

2.11 Other Standards Related to Safety and Cybersecurity

Some standards briefly outlined in this section were still in the process of being adopted or were still under development when this book went to press.

2.11.1 ISO/IEC 5888

ISO/IEC 5888[7] is intended to describe procedures for the development of cybersecurity requirements for connected vehicle components and suitable criteria for evaluating these. The standard follows and corresponds to ISO/IEC 15408, better known as *Common Criteria*.

[7] The standard is currently under development and its status can be viewed at https://www.iso.org/standard/81805.html.

2.11.2 ISO/TR 4804

The ISO/TR 4804 [17] standard describes the path to the development and validation of automated driving systems based on fundamental safety and cybersecurity principles derived from relevant publications. Safety and cybersecurity-by-design, as well as verification and validation methods for automated driving systems, are considered, with a focus on vehicles with level 3 and level 4 features. In addition, cybersecurity considerations are presented along with the safety goals for automated driving systems.

2.11.3 ISO PAS 8800

The ISO PAS 8800[8] standard, which is currently still under development, defines appropriate safety-related principles, methods, characteristics, and risk-influencing factors of artificial intelligence (AI) in relation to its use in motor vehicles. This is in accordance with the requirements of the ISO 26262 and ISO 21448 standards, and is also intended to harmonize the concepts of ISO/TR 4804 and ISO 21448. The standard covers all development and deployment phases. It describes the derivation of safety requirements, the consideration of data quality and completeness, architectural measures for error control, tools to support AI, and verification and validation techniques to ensure the overall safety of the system. It also provides evidence for formal verification.

2.11.4 ISO/IEC TR 5469

The technical report deals with the characteristics, potential risks, methods, and processes for the application of artificial intelligence in relation to functional safety (in safety-critical operations) [18].

2.11.5 IEC 61508

IEC 61508 [19] is an internationally recognized series of standards for the functional safety of electrical, electronic, and programmable electronic systems. The standard defines a framework for the management of functional safety in the development of systems that minimize the risk of technical faults or accidental failures. The standard is a basic standard on which many industry-specific functional safety standards are based, such as ISO 26262 in the automotive sector.

[8] The standard is currently under development and its status can be viewed at https://www.iso.org/standard/83303.html.

References

1. ISO/SAE 21434:2021 Road vehicles—Cybersecurity engineering. Standard, International Organization for Standardization/Society of Automotive Engineers, Geneva, CH (2021). https://www.iso.org/standard/70918.html
2. IEC TS 62443-1-1:2009 Industrial communication networks—Network and system security—Part 1-1: Terminology, concepts and models. Standard, International Electrotechnical Commission (2009). https://webstore.iec.ch/publication/7029
3. IEC TS 62443-1-5:2023 Security for industrial automation and control systems—Part 1-5: Scheme for IEC 62443 security profiles. Standard, International Electrotechnical Commission (2023). https://webstore.iec.ch/publication/67461
4. IEC 62443-2-1:2010 Industrial communication networks—Network and system security—Part 2-1: Establishing an industrial automation and control system security program. Standard, International Electrotechnical Commission (2010). https://webstore.iec.ch/publication/7030
5. IEC TR 62443-2-3:2015 Security for industrial automation and control systems—Part 2-3: Patch management in the IACS environment. Standard, International Electrotechnical Commission (2016). https://webstore.iec.ch/publication/22811
6. IEC TR 62443-3-1:2009 Industrial communication networks—Network and system security—Part 3-1: Security technologies for industrial automation and control systems. Standard, International Electrotechnical Commission (2009). https://webstore.iec.ch/publication/7031
7. IEC 62443-3-2:2020 Security for industrial automation and control systems—Part 3-2: Security risk assessment for system design. Standard, International Electrotechnical Commission (2020). https://webstore.iec.ch/publication/30727
8. IEC 62443-3-3:2013 Industrial communication networks—Network and system security—Part 3-3: System security requirements and security levels. Standard, International Electrotechnical Commission (2013). https://webstore.iec.ch/publication/7033
9. IEC 62443-4-1:2018 Security for industrial automation and control systems—Part 4-1: Secure product development lifecycle requirements. Standard, International Electrotechnical Commission (2018). https://webstore.iec.ch/publication/33615
10. IEC 62443-4-2:2019 Security for industrial automation and control systems—Part 4-2: Technical security requirements for IACS components. Standard, International Electrotechnical Commission (2019). https://webstore.iec.ch/publication/34421
11. ISO/IEC 27001:2022 Information security, cybersecurity and privacy protection—Information security management systems—Requirements. Standard, International Organization for Standardization/International Electrotechnical Commission (2022). https://www.iso.org/standard/82875.html
12. ISO 26262:2018 Road vehicles—Functional safety, Part 1-12. Standard, International Organization for Standardization (2018)
13. ISO/PAS 5112:2022 Road vehicles—Guidelines for auditing cybersecurity engineering. Standard, International Organization for Standardization (2022). https://www.iso.org/standard/80840.html
14. ISO 19011:2018 Guidelines for auditing management systems. Standard, International Organization for Standardization (2018). https://www.iso.org/standard/70017.html
15. German Association of the Automotive Industry (VDA): VDA ISA Catalog Version 6.0.1 (2023). https://www.vda.de/en/topics/digitization/data/information-security
16. ENX: Introducing Vehicle Cybersecurity (VCS) Audit Scheme (2023). https://portal.enx.com/en-us/news/Introducing-Vehicle-Cybersecurity-VCS-Audit-scheme/
17. ISO/TR 4804:2020 Road vehicles—Safety and cybersecurity for automated driving systems—Design, verification and validation. Standard, International Organization for Standardization (2020). https://www.iso.org/standard/80363.html

18. ISO/IEC TR 5469:2024 Artificial intelligence—Functional safety and AI systems. International Electrotechnical Commission (2024). https://www.iso.org/standard/81283.html
19. IEC 61508-1:2010 Functional safety of electrical/electronic/programmable electronic safety-related systems—Part 1: General requirements. Standard, International Electrotechnical Commission (2010). https://webstore.iec.ch/publication/5515

Chapter 3
Integrated Management Systems and Integrated Risk Management

Abstract In this chapter, we examine the characteristics of various standards in cyber and information security and (cyber) risk management. We provide an overview of the risk management methods and systems of the previously considered standards, with the aim of assessing their fundamental suitability for integration into a unified (risk) management system. We do not delve into technical details or legal restrictions. We consider practical aspects and possible limitations in integrating such systems.

The previous two chapters provided an overview of the key cybersecurity legal norms and technical norms and standards for the manufacturing industry. It became clear that there are different regulatory and technical perspectives on the domains of cybersecurity, information security and risk management. While some standards, particularly those relating to management systems, focus on the overall organization, other standards focus on different aspects or sub-aspects. For example, the ISO/IEC 27001 standard is aimed at organizations and sets out the requirements for an effective management system, which can then be used to establish and improve information security in all areas and levels and for all aspects of an organization. It is not relevant to what purpose this organization pursues. If external influencing factors are disregarded for a moment, it depends solely on the will of the organization at which points the standard is implemented[1], e.g., in which organizational units, at which locations or for which business processes. For example, information security can be established in the area of product development, in product manufacturing, in the service or operation of such products themselves, or anywhere else. The abstract nature of the standard—the body of the standard itself extends to only around ten pages of text—allows it to be implemented regardless of size, sector or corporate culture.

On the other hand, there are very specific standards with precisely described procedures, such as ISO/SAE 21434 and the TARA defined therein. The focus here is on the manufacturing industry, or, more precisely, the manufacturers of road vehicles. The more industry-specific characteristics, requirements, or objectives are incorporated into such a standard, the more tailored the result can be. This often leads to easier implementation in the organizations concerned because the special character-

[1] The standard refers to this as the scope of application.

R. do Carmo and A. Schlensog, *Automotive Threat Analysis and Risk Assessment in Practice*, https://doi.org/10.1007/978-3-662-69614-9_3

istics have already been considered in advance. An associated disadvantage may be that such a standard is difficult or impossible to apply to other sectors.

If an organization is faced with the challenge of having to apply several standards for regulatory reasons or due to market requirements, it should be examined whether these should be implemented and operated in an integrated manner. A look at practice shows that organizations often avoid integration because it increases project complexity. It is said that this requires an increased use of resources, that the project then has a longer duration, or that certain parts of the organization are already too busy anyway. Integration can help to reduce the resources required for operations.

Looking at standards that define requirements for management systems (e.g., ISO/IEC 27001 or ISO 9001), it is noticeable that the majority of these are very brief in terms of their scope. Compared to technical standards, management system standards typically contain rather abstract requirements. To enable simple integration of management systems from different disciplines, the High-Level Structure (HLS) was developed, which has been applicable to all ISO management system standards since 2012. In May 2021, the HLS became the "harmonized structure", which entails slight adjustments. The new structure will be applied in future editions of the standards.

In the next section, we take a closer look at the standards already considered and focus on a central aspect, risk management. To this end, we list the risk management standards and the main risk management methods from the standards presented above to assess their fundamental suitability for integration into existing risk management systems.

3.1 Overview of Risk Management Methods and Systems

The legal standards considered in Chap. 2 are of European law rank or, in some cases, national law in their implementation and do not generally address technical issues or details. The risk management-specific requirements of these legal standards do not contain any restrictive or incompatible aspects. For example, the assessment and appropriate and effective treatment of risks are required, but the design of the processes to be used for this is not described in more detail or restricted. The design is left to the parties concerned or is concluded through the application of standards. In the following, we list the key requirements for risk management from the technical and management system standards listed in Chap. 2 and consider the compatibility of these requirements with each other.

3.1.1 ISO/IEC 27001

ISO/IEC 27001 [1] essentially requires the definition and implementation of a risk management process and, within this, the consideration of probabilities and damage (impact) per asset or asset group. An appropriate and comprehensible risk acceptance level must be defined. The minimum protection goals to be considered are availability, integrity, and confidentiality.

3.1.2 BSI-Standard 200-1

BSI-Standard 200-1 [2] requires the definition and implementation of a risk management process. Predefined risks and lists of measures per asset group are to be considered via the compendium, which serve to achieve the basic level of protection. If the need for protection is increased, a more extensive risk analysis and treatment must be carried out. A method compliant with ISO/IEC 27001 can be used here. The minimum protection objectives to be considered are availability, integrity, and confidentiality. BSI-Standard 200-3 [3] provides extensive guidance on the design, but this does not restrict the applicable method.

3.1.3 ISO/IEC 27005

A generic risk management process is described at ISO/IEC 27005 [4], which can be used within the implementation of ISO/IEC 27001. This means that the system is fully compatible or does not have any restrictive requirements.

3.1.4 ISO 31000

The ISO 31000 [5] standard describes a generic risk management process and not a specific one for the management of information security risks. However, the process described can be applied without problems, e.g., within ISO/IEC 27001, as it is basically the same as the process described in ISO/IEC 27005. There are no conflicts of objectives or incompatibilities with the aforementioned standards.

3.1.5 ISO 22301

One of the first steps in introducing a business continuity management system is to carry out a business impact analysis (BIA) [6]. This examines the impact of the failure of the processes under consideration on business operations. The focus is on the protection goal of availability. In addition, a risk analysis is used to examine the risks to which assets or asset groups that are required to carry out the previously examined business processes are exposed. The focus here is also on availability. The overall process, consisting of the BIA and subsequent risk analysis, can be integrated into a risk management system in accordance with ISO/IEC 27001, for example. The BIA, which is not formally provided for in ISO/IEC 27001, is suitable for gaining a more profound understanding of the business processes to be protected and the associated information. This analysis data can be used to define the protection requirements (criticality) of the information to be protected more precisely. A common risk management process can then be used in the subsequent risk analysis. Only the protection objectives that are required for the respective assets or asset groups then need to be considered.

3.1.6 TISAX®

TISAX® (Trusted Information Security Assessment Exchange) was created to establish a uniform methodology for determining the information security level of an organization. The requirements for risk management correspond to the requirements of ISO/IEC 27001 and can therefore be easily integrated into a common system. The minimum protection goals to be considered are availability, integrity, and confidentiality. Any necessary consideration of data protection and the resulting extension of the protection objectives does not contradict the fundamental requirements of a risk management system.

3.1.7 IEC 62443

There are several standards to consider, which also place different requirements on risk management due to their different areas of application.

IEC 62443-2-1 addresses the requirements for a CSMS[2] that is aimed at the organization. This standard was developed with the aim of making the requirements consistent with ISO/IEC 27001. Assets or asset groups are only described in very abstract categories as an asset to be protected, such as a technical or logical object. The systems specifically covered by the series of standards are industrial monitoring and control systems, which means that there are no restrictions or even different views on any of the aforementioned standards[3]. The protection goals (objectives) can be freely supplemented; in principle, confidentiality, integrity, and availability are considered. However, the goals are typically prioritized differently for each area of application. The risk management process, which essentially consists of the

[2] The concept of CSMS was removed from IEC 62443 in the second edition of IEC 62443-2-1 (August 2024). The standard now requires coordinating the security program with the organization's ISMS or introducing a suitable process if no ISMS exists.

[3] In addition to ISO/IEC 27001, ISO/IEC 27019 defines special requirements in the area of energy supply.

assessment and treatment of risks, corresponds to the systems mentioned above. The organization is free to define the risk tolerance[4] itself.

IEC 62443-3-2, on the other hand, addresses the cybersecurity of industrial automation and control systems and their impact on the organization. The standard uses a process model based on the application of the concept of security zones and conduits. The risk management system defined in this way is more comprehensive and more precisely defined than in the previously considered standards. However, it is based on familiar elements such as asset groups, definitions of protection requirements (target security level), or gross and net risk assessments[5]. Treatment can be by transfer, acceptance, or mitigation. The organization determines the tolerable risk level itself. The risk value is determined by the assessment of probability and impact.

Even if the procedure initially appears to differ greatly from the simple models of ISO/IEC 27001, ISO/IEC 27005, or ISO 31000, the general process is comparable. In principle, the models can be designed and implemented consistently with each other.

3.1.8 ISO/SAE 21434

The ISO/SAE 21434 [7] standard describes a cybersecurity management system and the requirements for the system. Only the consideration of cybersecurity risks to road vehicles is required. The standard describes a management system from a governance perspective, but does not consider the entire organization and its context, but focuses on road vehicles and their components for their entire life cycle. The risk management system requires the identification of threat scenarios at asset level, the assessment of the impact, the analysis of attack paths, the assessment of the feasibility of attacks, and the determination of the risk value. A suitable procedure, known as TARA (threat analysis and risk assessment), is also defined for this purpose. The typically used security objectives (cybersecurity properties) are confidentiality, availability, and integrity, which can be supplemented by further objectives. If a risk is to be retained (risk retaining), cybersecurity claims must be specified to justify the appropriateness of the approach. This corresponds to the concept of risk acceptance. It is also possible to omit further treatment for threat scenarios with a risk value of 1.

---

[4] In ISO/IEC 27001, for example, referred to as risk acceptance.

[5] Gross risk corresponds to the unmitigated risk, the net risk corresponds to the residual risk.

As with IEC 62443-3-2, the procedure differs greatly from the simple models, but the basic concept is also comparable in this case. This means that the risk management procedures can be designed and implemented consistently with each other. It should be noted that a risk analysis, as required by IEC 62443-3-2, can be implemented using a TARA. In organizations that manufacture products for the road vehicle sector, but also other products, it is advisable to use the TARA concept and not apply different procedures. In the best-case scenario, this procedure and a company-wide CSMS or ISMS should be made consistent and operated in an integrated manner.

ISO/SAE 21434 focuses primarily on cybersecurity engineering from the perspective of a single item or component, which can be a part of a vehicle, such as an electronic control unit (ECU), or the vehicle as a whole, i.e., analyzing its E/E architecture. Regarding the production of an item, i.e., the production of an embedded device by a supplier, or the production of the entire vehicle, there are basically two main objectives that the standard addresses in its chapter *12 Production* when addressing the manufacturing and post-development phases. These objectives are:

1. the creation of cybersecurity requirements for post-development, and
2. the prevention of the introduction of vulnerabilities during production.

ISO/SAE 21434 requires that a production control plan be in place that includes cybersecurity controls to prevent unauthorized modification of the item during production. Examples of controls, according to the standard, can be physical controls that prevent physical access to production servers containing software, or logical controls that use cryptographic techniques and access controls. To achieve these objectives, a comprehensive risk management of the production process must be in place.

A manufacturing plant can be viewed as a complex environment of both information technology (IT) and operational technology (OT) systems.

To facilitate understanding and integration of risk management systems, we introduce a comparison of key terms from the standards in Table 3.1.

Table 3.1 Comparison of key terms from different standards for risk management systems

ISO/IEC 27001[a] [8]	**BSI-Standard 200-1**[b] [3]	**ISO/IEC 27005**[a] [8]	**ISO 31000**[5]	**ISO 22301**[c] [9]	**TISAX®**[d]	**IEC 62443**[d]	**ISO/SAE 21434** [7]
Risk	Risk	Risk	Risk	Risk	Information security risk [10]	Risk [11]	Risk
Risk assessment	Risk assessment	Risk determination	Risk assessment	Risk assessment	Risk assessment [12]	Risk assessment [11]	Threat analysis and risk assessment
Risk identification	Drawing up of a threat overview	Risk identification	Risk identification	Risk identification	Risk identification [13]	Identify threats[e] [14]	Threat scenario identification[e]
Risk analysis	Risk assessment	Risk analysis	Risk analysis	Risk analysis	Risk analysis [13]	Risk assessment [11]	Threat analysis and risk assessment
Risk evaluation	Risk evaluation	Risk evaluation	Risk evaluation	Risk evaluation	Risk evaluation [13]	Determine unmitigated cyber security risk [14]	Risk value determination
Consequence	Extent of damage, effects of damage	Risk identification	Risk analysis	Consequence	Impact, potential damage [10]	Consequence [11, 14], impact [14]	Impact
Likelihood	Frequency of occurrence	Risk analysis	Risk evaluation	Likelihood	Probability of occurrence, frequency of occurrence [13]	Likelihood [14]	Attack feasibility

(Continued)

Table 3.1 (Continued)

ISO/IEC 27001[a] [8]	**BSI-Standard 200-1**[b] [3]	**ISO/IEC 27005**[a] [8]	**ISO 31000**[5]	**ISO 22301**[c] [9]	**TISAX®**[d]	**IEC 62443**[d]	**ISO/SAE 21434** [7]
Risk acceptance criteria	Risk acceptance criteria	Risk evaluation	Risk criteria	Not a precise term exists, but the concept is applied through the process of risk evaluation	Threshold [13]	Risk tolerance level [11], tolerable risk [14]	Not a precise term exists, but the concept is applied in *[PM-06-08]* of [7]
Risk acceptance	Risk acceptance	Risk retention	Retaining the risk	Risk acceptance	Risk acceptance [13]	Accept the risk [11]	Retaining the risk
Risk treatment	Risk treatment	Risk treatment	Risk treatment	Risk treatment	Risk treatment [13]	Risk response [11], compare unmitigated risk with tolerable risk [14]	Risk treatment decision
-[f]	Risk avoidance	Risk avoidance	Risk avoidance	Avoiding the risk	Risk avoidance [13]	Design the risk out [11]	Avoiding the risk
-[f]	Risk reduction, risk modification	Risk modification	Changing the likelihood, changing the consequences	Risk mitigation, risk reduction	Risk mitigation [13]	Reduce the risk [11], mitigate the risk [14]	Reducing the risk
-[f]	Risk transfer, risk sharing	Risk sharing	Sharing the risk	Risk sharing	Risk transfer [13]	Transfer or share the risk [11]	Sharing the risk

(Continued)

Table 3.1 (Continued)

ISO/IEC 27001[a] [8]	**BSI-Standard 200-1**[b] [3]	**ISO/IEC 27005**[a] [8]	**ISO 31000**[5]	**ISO 22301**[c] [9]	**TISAX®**[d]	**IEC 62443**[d]	**ISO/SAE 21434** [7]
Residual information security risk	Residual risk, remaining risk	Residual risk	Remaining risk	Residual risk	Residual risk [13]	Residual risk [11]	–
Level of risk	Risk category	Level of risk	Level of risk	Level of risk	Level of risk [13]	Level of risk [11]	Risk value
Asset	Asset	Asset	Asset	Asset	Information asset [13]	Asset [11]	Asset
System	–	–	–	–	-	System under consideration [14]	Item, component (not same meaning)
Control	Safeguard	Control	Control	Control	Measure [13]	Countermeasure [11]	Cybersecurity control
Security objective	Security objective	Security objective	-	-	Protection objective [13]	Security objective [11]	Cybersecurity property

[a] ISO/IEC 27001 and ISO/IEC 27005 refer to the definitions from ISO/IEC 27000 in Chap. 3
[b] Risk management is defined by BSI-Standard 200-3
[c] ISO 22301 refers to the definitions from ISO 22300 in Chap. 3
[d] Citations come from various sources; therefore, they are placed by the term
[e] Less abstract but the same concept is used
[f] Term is not used directly in ISO/IEC 27001; reference is made to the concept of ISO 31000

3.2 Harmonization and Integration of Management Systems

In practice, management systems are often operated side by side and independently of each other. The individual management systems differ in terms of their structural and procedural organization. The management of the systems reports to different places in the organization, and the system also frequently has an individual area of application. This leads to significantly higher costs for the organization, as structures, processes, procedures, reporting channels, and operations are highly redundant. In the worst case, aspects of the systems are incompatible with each other, or individual objectives directly or indirectly contradict each other.

An integrated management system (IMS) creates order. It combines the different management systems of a company into one system with a standardized structure. In this way, different domains such as quality management, occupational health and safety, and information security can be integrated into a common organizational and process structure. Integrated management systems are rarely implemented directly in an organization. Usually, isolated systems are combined at a later stage.

Ideally, work is done on common information and data basis, which leads to consistent results. In this way, risk management could be carried out in an integrated manner and thus lead to comparable, coherent, and consistent results across different domains and levels of abstraction. Negative effects can be minimized or even excluded.

As useful as the subsequent integration of isolated systems into an integrated management system is, implementation can be difficult. An integrated management system is more complex in its structure and implementation than a single, isolated system. Therefore, planning and operation require greater foresight and experience. Even with the full support of top management, organizations find it difficult to dismantle the structures that then become unnecessary. Established organizational forms and power structures slow down such projects or lead to the project coming to a standstill or even being abandoned.

Due to the harmonization of ISO management system standards, these are theoretically easy to integrate. What needs to be considered, and what aspects arise in practice when integrating such systems and, in particular, risk management?

Different areas of application within an integrated management system do not usually create contradictions or problems. The complexity of the system can increase with the number of different application areas. The aim should therefore be to minimize the number. Each management system serves a specific area and pursues different objectives. These often abstract objectives should be harmonized. As a rule, there are no contradictions at this level of consideration. What initially appears to be a conflict of objectives can be easily resolved.

For example, work safety as an objective of a management system, i.e., the protection of life and limb, is naturally prioritized higher than, for example, the availability objective of another system.

An example that seems contrived is an emergency stop button. This may be necessary due to work and operational safety requirements. At the same time, an emergency stop button represents a hazard. Depending on the design, triggering it can shut down entire production systems or server rooms. If the emergency stop button is protected against accidental activation by a flap, the potential risk is reduced. The contradiction is resolved.

The following example from the area of data protection and data economy is slightly less direct. The operation of IT requires the ability to evaluate metadata or retain log data to be able to analyze and clarify incidents retrospectively. This makes it easier, for example, to better understand and comprehend complex attacks that have been prepared for a long time and to find effective measures against attacks. Data protection aims to ensure that such data is not collected or stored in the first place. Assuming that applicable law is not violated, this issue can be examined as part of an analysis.

Opportunities and risks, or costs and benefits, are weighed up against each other. From the perspective of a management system, this represents active risk management. The issue is considered from different angles, effects are evaluated, and possible courses of action are weighed up against each other. When integrating risk management systems or methods, care must be taken to ensure that definitions of, e.g., damage categories, threshold values or metrics, and methods used are aligned or, better yet, derived directly from each other. Risks are often assessed at different company levels and must then be combined to form a coherent picture of the company's overall risk.

For example, if risks are assigned to an individual product, they can be aggregated to constitute an overall contribution for the producing business unit. In turn, the contribution in relation to the organization's overall business can be derived from this. A further difficulty lies in the fact that risk assessments are frequently not "exactly", i.e., cardinally, but only "evaluated" in the form of rankings, i.e., ordinal.

When it comes to cybersecurity risk management in production facilities, i.e. risk management for IT and OT systems, there are two large, well-established families of international standards, namely the ISO/IEC 27000-series [8, 1, 5] for IT systems and the IEC 62443 [11, 16] for OT systems.

For larger companies or manufacturing facilities, there is a good chance that the company and manufacturing facilities are certified to ISO/IEC 27001 and/or IEC 62443. Certification to one of these standards may be a valid argument for achieving the objectives required by ISO/SAE 21434. However, it is possible that a complete certification for production facilities is not available. For example, a company may be certified for its internal IT system, but not for the special case of production facilities, such as servers and machines dedicated to production. For this reason, it may make sense to include risk management for production facilities in the scope of a TARA according to ISO/SAE 21434. It is a challenge to create a uniform understanding and awareness of the various risk management approaches throughout the company. In this book, we attempt to create this understanding of the TARA method. We want to take advantage of the well-established standards for IT and OT and apply some of their methodologies, such as threats and controls, to our ISO/SAE 21434-compliant TARA. The dilemma of whether to apply

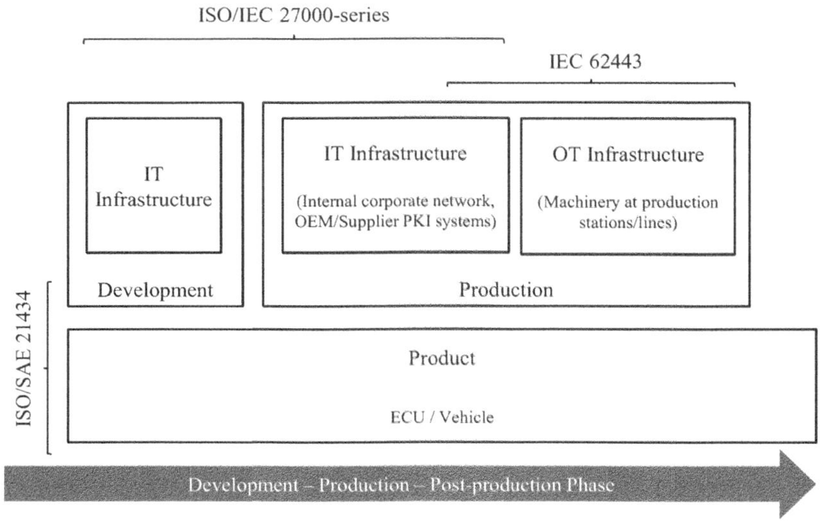

Fig. 3.1 High-level coverage of the ISO/SAE 21434 during the vehicle/product life-cycle and the standards for IT and OT

ISO/IEC 27001 or IEC 62443 in a mixed IT/OT system is not new to the industry[6]. An approach for combining these two standards has been formalized in [17], and we take inspiration from it for application to automotive cybersecurity, apart from our professional experience with similar TARA projects. In Fig. 3.1 we show the high-level overlap between the different standards for the different phases of the product life cycle.

3.3 An ISO/SAE 21434 TARA for IT/OT Systems

The objective of the ISO/SAE 21434 with regard to production is to protect the item (e.g., ECU or vehicle) against vulnerabilities introduced during its production and in the post-development phase.

For example, it shall be ensured that no malicious software can be introduced in the ECU by a malicious operator or because of an unprotected service exposed to the Internet. At the post-development phase, for example, it has to be ensured that the back-end servers having authentication material for the ECUs cannot be compromised getting access to the ECUs in the field. The combination of the IT and OT with ISO/SAE 21434 derives from the fact that a threat analysis and risk

[6] The concept of CSMS was removed from IEC 62443 in the second edition of IEC 62443-2-1 (August 2024). The standard now requires coordinating the security program with the organization's ISMS or introducing a suitable process if no ISMS exists.

assessment at production level has to identify risks for the item during production and provide controls to mitigate them (of course, assuming mitigation as decision treatment).

We take advantage of the controls and experience provided by standards ISO/IEC 27000-series and IEC 62443 and combine them to the TARA. [17] follows a similar approach for combining these standards. We highlight that a TARA following the ISO/SAE 21434 standard aims at protecting the ECU or the vehicle. Combining the controls and requirements from the standards ISO/IEC 27000-series and IEC 62443 does not mean compliance to these standards. However, assuming risks found due to an unprotected environment, following controls and requirements from well established standards in the areas of IT and OT can be extremely beneficial.

References

1. ISO/IEC 27001:2022 Information security, cybersecurity and privacy protection—Information security management systems—Requirements. Standard, International Organization for Standardization / International Electrotechnical Commission (2022). https://www.iso.org/standard/82875.html
2. BSI Standard 200-1: Information Security Management Systems (ISMS). Standard, Bundesamt für Sicherheit in der Informationstechnik (2017). https://www.bsi.bund.de/dok/10027834
3. BSI Standard 200-3: Risk Analysis based on IT Grundschutz. Standard, Bundesamt für Sicherheit in der Informationstechnik (2017). https://www.bsi.bund.de/dok/10027834
4. ISO/IEC 27005:2022 Information security, cybersecurity and privacy protection—Guidance on managing information security risks. Standard, International Organization for Standardization / International Electrotechnical Commission (2022). https://www.iso.org/standard/80585.html
5. ISO 31000:2018 Risk management—Guidelines. Standard, International Organization for Standardization (2018). https://www.iso.org/standard/65694.html
6. ISO 22301:2019 Security and resilience—Business continuity management systems—Requirements. Standard, International Organization for Standardization (2019). https://www.iso.org/standard/75106.html
7. ISO/SAE 21434:2021 Road vehicles—Cybersecurity engineering. Standard, International Organization for Standardization/Society of Automotive Engineers, Geneva, CH (2021). https://www.iso.org/standard/70918.html
8. ISO/IEC 27000:2018 information technology—security techniques—information security management systems—overview and vocabulary. Standard, International Organization for Standardization / International Electrotechnical Commission (2018). https://www.iso.org/standard/73906.html
9. ISO 22301:2021 Security and resilience—Vocabulary. Standard, International Organization for Standardization (2021). https://www.iso.org/standard/77008.html
10. German Association of the Automotive Industry (VDA): VDA ISA Catalog Version 6.0.1 (2023). https://www.vda.de/en/topics/digitization/data/information-security
11. IEC TS 62443-1-1:2009 Industrial communication networks—Network and system security—Part 1-1: Terminology, concepts and models. Standard, International Electrotechnical Commission (2009). https://webstore.iec.ch/publication/7029
12. ENX: TISAX® Participant Handbook Version 2.7. Standard, ENX (2023). https://www.enx.com/handbook/tisax-participant-handbook.html

13. VDA: Information Security Risk Management—Protection objectives: confidentiality, integrity, availability—Version 1.1. White paper, German Association of the Automotive Industry (VDA) (2021). https://www.vda.de/en/news/publications/publication/white-paper--information-security-risk-management-
14. IEC 62443-3-2:2020 Security for industrial automation and control systems—Part 3-2: Security risk assessment for system design. Standard, International Electrotechnical Commission (2020). https://webstore.iec.ch/publication/30727
15. ISO/IEC 27002:2022 Information security, cybersecurity and privacy protection—Information security controls. Standard, International Organization for Standardization / International Electrotechnical Commission (2022). https://www.iso.org/standard/75652.html
16. IEC 62443-2-1:2010 Industrial communication networks—Network and system security—Part 2-1: Establishing an industrial automation and control system security program. Standard, International Electrotechnical Commission (2010). https://webstore.iec.ch/publication/7030
17. International Society of Automation: Applying ISO/IEC 27001/2 and the ISA/IEC 62443 Series for Operational Technology Environments (2021). https://www.isa.org/news-press-releases/2021/july/new-white-paper-applying-iso-iec-27001-2-and-the-i

Chapter 4
Threat Analysis and Risk Assessment

Abstract In this chapter, we provide a more in-depth introduction to the threat analysis and risk assessment process. We highlight the significance of TARA within the automotive industry, explaining its purpose and necessity. We also describe when a TARA needs to be performed in the vehicle life cycle, based on our professional experience and the regulatory framework. We also discuss the duration and level of technical detail of a TARA project according to the life cycle phase. The overall TARA process covered in this book is described in the following chapters.

A Threat Analysis and Risk Assessment (TARA) is a systematic, analytical analysis designed to identify potential threats that may pose risks to a given system. The primary objective of a TARA is to assess these risks and formulate effective countermeasures to mitigate them to an acceptable level. A TARA is a highly technical and challenging process in which the system to be assessed is carefully and thoroughly analyzed. A TARA requires intimate knowledge of the system to be assessed, combined with cybersecurity specific knowledge to identify every possible threat. Performing a TARA combines domain-specific knowledge of the system under evaluation with specialized cybersecurity expertise. This integration allows for the exploration of different attack vectors and vulnerabilities that could be exploited by adversaries.

A TARA is a generic process that can be applied to any system in any domain. For instance, in [1] the authors make use of concepts from ISO/SAE 21434 for applying TARA to medical devices. Especially for the automotive sector, the concept of TARA has existed for many years, with earlier well-known approaches such as EVITA [2] and HEAVENS [3]. Other popular approaches to TARA, not explicitly designed for the automotive industry, but also used in the automotive industry, include ETSI TVRA [4], PASTA [5], the well-known international standard for information security management ISO/IEC 27000-series [6], and others. With the emergence of regulations and standards in recent years, such as UN Regulation No. 155 and the ISO/SAE 21434 standard, the TARA process has been heavily leveraged and standardized for the automotive industry.

Supplementary Information The online version contains supplementary material available at https://doi.org/10.1007/978-3-662-69614-9_4.

R. do Carmo and A. Schlensog, *Automotive Threat Analysis and Risk Assessment in Practice*, https://doi.org/10.1007/978-3-662-69614-9_4

The primary focus of TARA according to ISO/SAE 21434 [7], which this book follows, is the vehicle and its internal components. However, the automotive sector is a complex ecosystem of vehicles, off-vehicle devices, IT systems, manufacturing and OT systems working together. For this reason, we address two major groups of systems in this book. The TARA methodology used in all cases is that described in the ISO/SAE 21434 standard. First, we analyze embedded devices, i.e., electronic control units (ECUs) inside the vehicle, as well as vehicle-level E/E architectures. These types of TARAs are, in our experience, the most requested analyzes, as they are explicitly within the scope of the ISO/SAE 21434 standard. Second, we analyze off-vehicle systems, such as manufacturing plants and back-end systems, because they are also part of a comprehensive cybersecurity posture.

What is Not a TARA?
It is important to clarify, especially for readers unfamiliar with cybersecurity risk management and automotive TARA, that a TARA should not be confused with a hands-on security test such as a penetration test. While a penetration test involves the actual execution of attacks to identify vulnerabilities, a TARA serves a different purpose. A TARA focuses on analyzing potential threats and assessing the associated risks in a systematic and structured manner. It aims to identify and understand the various threats that a system may face, assess their potential impact, and propose strategies to mitigate the risk. Unlike a hands-on security test, a TARA does not involve actively testing or attacking the system. In addition, it is important to note that a TARA cannot provide definitive proof that a system is secure. It is an assessment that helps uncover vulnerabilities and weaknesses, but it does not guarantee absolute security. Rather, it provides valuable insight into potential risks and helps make informed decisions to increase the overall cybersecurity posture.

4.1 Why is It Necessary?

There are good reasons for doing threat analysis and risk assessment in the automotive industry and in general. First, *risk management* is critical to ensuring the production of secure vehicles with the ultimate goal of protecting human life. The automotive industry recognizes that cybersecurity is as important as safety in ensuring the well-being of customers and road users. In addition, the perception of secure products is today a fundamental aspect of product quality. Customers expect a product to be secure as much as they expect the product to be safe and function as intended. In addition, manufacturers themselves have a vested interest in protecting their business operations from various threats. Potential risks such as data leakage and denial-of-service attacks can have serious consequences, including reputational damage, financial loss, and loss of customer confidence. By proactively identifying and mitigating these risks, manufacturers can protect their business continuity and maintain a competitive edge in the market.

Second, *regulatory compliance* is of paramount importance in the automotive industry. Industry-specific standards and regulations, such as UN Regulation No. 155 and ISO/SAE 21434, provide guidelines and requirements for ensuring cybersecurity in vehicles. Performing TARA becomes a necessity to comply with these regulations and meet industry and legal mandates. By integrating TARA into the product development life cycle, manufacturers can demonstrate compliance with regulatory frameworks and reinforce their commitment to cybersecurity governance.

Finally, systematically assessing a product's risks helps ensure that the product is developed in a cost-effective manner. By identifying and addressing potential vulnerabilities and threats early in the development process, manufacturers can avoid costly redesigns or fixes later.

4.2 When Does It Take Place in the Vehicle Life Cycle?

According to ISO/SAE 21434, the TARA steps, or methods, are generic modules that can be invoked systematically, and from any point in the life cycle of an item or component [7]. We combine the flexibility of the standard with our experience in the industry, and provide an overview of the phases in the life cycle where performing TARA is a good idea. We describe different levels of abstraction for TARA for the different phases.

There are four key phases in the vehicle life cycle where performing TARA is essential, considering the aspects of building and maintaining a secure product, developing the product cost-effectively, and complying with regulations and standards. The different types of TARA align with the life cycle phases outlined in the ISO/SAE 21434 standard. It is important to note that the different levels of abstraction of TARA follow a practical approach and can be used as guidance. Differentiating TARA at various phases has several benefits. First, it allows for the optimization of the TARA process itself. For instance, an initial TARA conducted during the concept phase may be relatively short, as detailed analysis of attack paths might not be feasible at that stage. However, it still serves the purpose of identifying high-level risks and establishing a foundation for subsequent assessments. Second, performing TARA at different phases helps maintain cost efficiency during product development. By conducting a comprehensive TARA during the development phase, new controls that were not initially planned for can be incorporated into the product design. This approach prevents the need for extensive redesign and expensive modifications at later stages. If the TARA is performed at the end of the development phase, the engineering costs of redesigning the product to mitigate risks will be much higher. The different TARA approaches are illustrated by the different phases of a typical vehicle life cycle in Fig. 4.1.

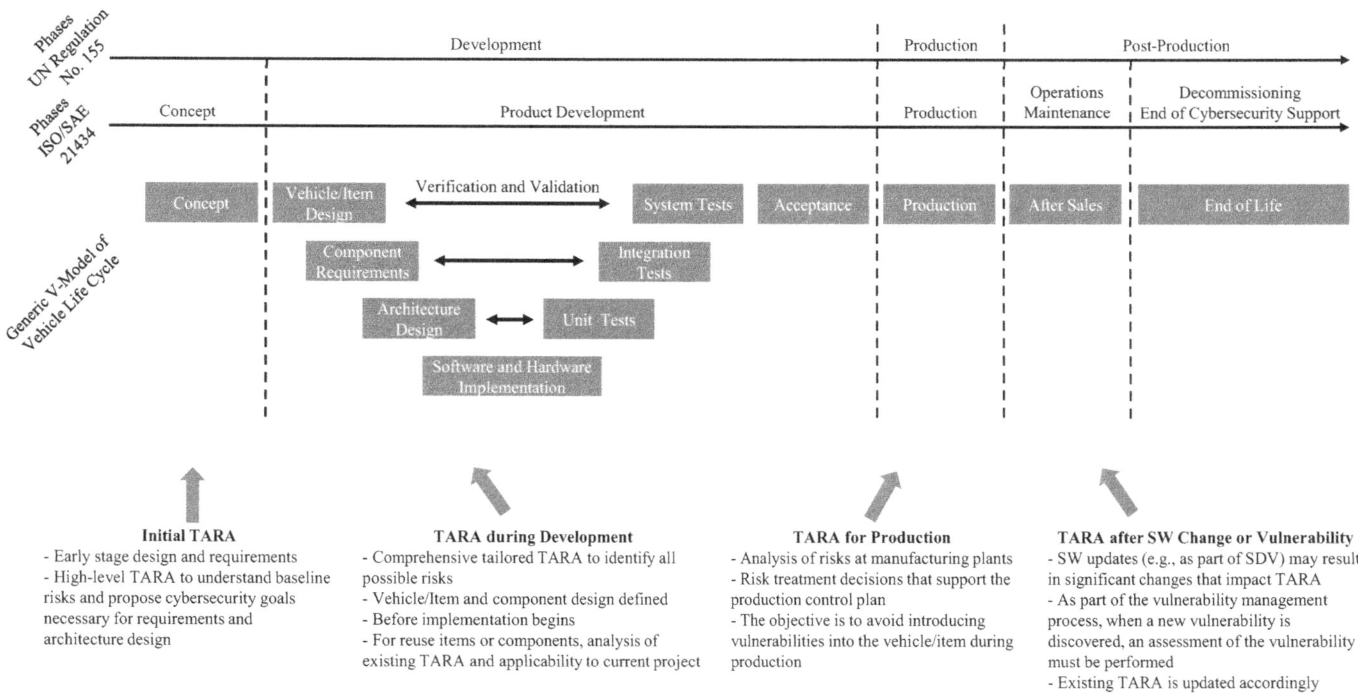

Fig. 4.1 TARA in the different phases of the vehicle life cycle

4.2.1 Initial TARA

The primary objective of the early or initial TARA is to support the concept phase by identifying high-level risks. At this stage, when only the operational context of the item is known, or assumed, it is not feasible to perform a detailed technical analysis. However, it is essential to identify and address high-level risks to lay the foundation for the cybersecurity concept and product design. For example, consider a scenario where external connectivity is planned for the item, but the specific connectivity system, such as Wi-Fi or Bluetooth, has not been determined. In this case, an initial TARA can analyze both connectivity systems and provide high-level cybersecurity goals and cybersecurity claims. These findings support the cybersecurity concept of the item and help establish the necessary security requirements. Then, during the design process, the engineering team may decide to include only Wi-Fi connectivity based on the TARA findings. This decision could be driven by cost-effectiveness, as achieving the desired cybersecurity goals may not be feasible with Bluetooth technology. It is important to note that this example is simplified to illustrate the benefits of early TARA and does not imply that one connectivity system is inherently superior or inferior to the other. By conducting an early-stage TARA, cybersecurity teams can proactively identify potential risks and make informed decisions about the selection of cybersecurity controls and technologies. This approach ensures that cybersecurity considerations are integrated into the design of the item from the earliest stages, facilitating a cost-effective and robust cybersecurity posture throughout the development process.

4.2.2 TARA During Development

The TARA during the development phases represents the typical exhaustive TARA for the item. This TARA requires meticulous technical rigor. The development phase itself is inherently complex and can be further subdivided into various sub-phases, each of which introduces potential changes and additional requirements to the design of the item. During the development phase, design elements of the item may undergo minor changes, new requirements may emerge, and vulnerabilities may be discovered through vulnerability analysis or other means. As a result, the TARA becomes a living document that evolves with the development phase. The cybersecurity team has the flexibility to determine the appropriate time to initiate the TARA during the development phase. Once initiated, the cybersecurity team assumes responsibility for regularly updating the TARA to reflect any changes in the item's design, new vulnerabilities, or the evolving threat landscape.

4.2.3 TARA for Production

The production TARA is a slightly different type of TARA than the item development TARA. The purpose of the TARA for production is not to analyze the item being produced, but to analyze the risks to the item in the production and post-production processes. During the production phase, the design of the item, whether it is a vehicle or a part, is already finalized and cannot be changed. The production phase TARA considers the production environment. For example, it ensures that no malicious software can be introduced into the ECU by a malicious operator or through an unprotected service exposed to the Internet. In the post-development phase, for example, it must be ensured that the back-end servers that hold authentication material for the ECUs cannot be compromised to gain access to the ECUs in the field. The combination of IT and OT with the methodology of the ISO/SAE 21434 stems from the fact that a threat analysis and risk assessment at the production level must identify risks to the item during production and provide controls to mitigate them (assuming, of course, that mitigation is the decision treatment).

4.2.4 TARA After SW Change or Vulnerability

The current era of Software Defined Vehicles (SDV), where vehicles have more and more software, with updatable core software features like modern cell phones, brings the discussion of how to handle cybersecurity management in a rapidly evolving environment. A software update, regardless of its importance, represents a change to the vehicle and therefore to the existing TARA. The cybersecurity team responsible for a system may need to analyze the impact of a planned software update on the cybersecurity of the system. One approach may be to link the planned software changes to an SBOM and provide the TARA team with the SBOM and a description of the functional changes introduced by the software update. The TARA can determine if there are increased or new risks due to the planned software changes. The project team can work on the increased or new risks identified before the software update is rolled out.

As new vulnerabilities are discovered in the field, the TARA may need to be updated. This process is an integral part of incident response and vulnerability management, as outlined in the ISO/SAE 21434 standard. This TARA is not an entirely new TARA for the item, as it was at the beginning of the concept and development phases. However, the impact of the vulnerability on the item, its assets, and threat scenarios must be analyzed with sufficient technical rigor. Based on the findings of the vulnerability analysis, changes may need to be made to the item or to the component to effectively address the identified risks. This could involve implementing additional cybersecurity controls, updating software or firmware, or applying patches and fixes.

4.3 Detail Level According to Available Information

Even when a TARA follows a well-defined, step-by-step approach with a clear methodology, there is a good deal of freedom regarding the level of abstraction of certain steps. We observe a recurring discussion and disagreement regarding the level of detail and the resulting duration and effort of a TARA project. In this section, we discuss three levels of abstraction for TARA that we describe with different durations depending on the needs of the project.

> The following approaches are only suggestions based on our experience to date. The suggested duration of the TARA serves as a rule of thumb, especially if the cybersecurity team does not have enough experience in performing TARA. The suggested duration is an order of magnitude. The *3-days TARA* may be performed in one day for some projects, and in five or ten days for others. The *30-days TARA* can be a 15-days TARA or a 40-days TARA, to give an example. It is important to emphasize that the term *"detail level"* should not be associated with a decrease in technical quality and rigor. However, it should be possible for the cybersecurity team to think critically about certain steps of the TARA that do not provide a better result in the context of the TARA project, regardless of the amount of effort put into them. Especially in the larger TARAs, we observe that very detailed TARAs, for example, by analyzing every single communication message at every single interface, do not provide a better result (i.e., controls, goals) than less detailed TARAs. It is essential to understand that TARA is not a single event, but rather an ongoing process that should be carried out at various stages throughout the item's life cycle. This approach allows for necessary modifications and improvements as the item, its operational environment, and the cybersecurity threat landscape continue to change. As such, it is more advantageous to have a TARA process that is both manageable and repeatable, allowing for regular updates, as opposed to a one-off, overly detailed analysis.

4.3.1 "The 3-Days TARA"—Initial TARA

The first TARA at the very beginning of a project is a rather short TARA. Suppose the item is an ECU for a vehicle. In the concept phase of the item, the rough architecture of the item is probably defined. It is already known what functions the ECU has to provide, and there is some idea about the communication interfaces present on the ECU. If the target vehicle for the ECU application is a new platform under development, chances are that the OEM's requirements for the ECU are not yet fully defined. The goal of a TARA at this point is to establish the baseline for cybersecurity goals, cybersecurity claims, and controls for high-level risks. For example, if the ECU has a direct connection to the Internet, the TARA will likely identify many more risks than an ECU that does not have an Internet connection and only has a communication interface to a non-safety CAN bus on the vehicle. A TARA at this level is short because the design of the item is not finalized, and therefore it is not possible to define the threat scenarios with a high level of technical detail. The cybersecurity team must analyze the current specifications of the item and work with the engineers to identify the relevant assets and define their cybersecurity properties. The threat identification and evaluation steps do not take as much time and effort as an advanced TARA. For example, the initial TARA can be based on a catalog of threats, and it can use the attack vector-based approach for threat evaluation, which is very fast because it is based on only one parameter.

4.3.2 "The 30-Days TARA"—TARA During Development and Reuse

This TARA is performed during the development of the item. At this stage, the operational environment of the item is clear, as are the requirements for the item. The goal of the TARA is to perform a rigorous technical analysis that identifies the cybersecurity risks, taking into account all the technical details of the item. The results of the TARA are likely to provide useful insight into the technical controls recommended for the item to mitigate the risks. If the TARA is a system-level or vehicle-level TARA, the results are likely to provide high-level controls for all components of the system and cybersecurity goals that need to be communicated to all component developers. Since the TARA must be performed in accordance with the development schedule of the item, the cybersecurity team must apply its technical criteria and decide when to stop when a technical level is reached that provides enough information that cannot be improved. This TARA may also be a reuse project. In a reuse project, a previous TARA may be available, so the effort may be less than performing a TARA from scratch. However, any differences in the item and its operational environment must be carefully analyzed, and not just rely on previous results.

We provide some examples of questions that the cybersecurity team can consider when thinking about the balance between level of detail and effort required:

- *Is it possible to create groups of assets?* Yes, and it even makes sense. Well-designed trust boundaries reduce the complexity of TARA and make it easier to focus on the essentials. A typical case of grouping assets is the analysis of communication interfaces and protocols. Instead of analyzing every single message as an asset, they can be grouped according to their functionality or even safety relevance. One of the goals of this TARA is to provide technical controls. The controls usually affect the whole interface, and therefore it is not relevant to know exactly which message is responsible for the high risk (such as braking message), but to know that there are, for example, safety-relevant messages that can be manipulated. Of course, there may be cases where there is a technical interest in understanding the details (e.g., considering applying SecOC only to certain messages to save computation). In such a case, the cybersecurity team can decide to leave the grouping and perform a deeper analysis later on only the relevant messages, or not to do the grouping at all.

- *Can the damage scenarios*[1] *be grouped?* We recommend grouping the damage scenarios as much as possible. This can save a lot of effort in the impact analysis, including comments and justifications, for each stakeholder and impact category of the damage scenarios, without causing any loss in the TARA. An essential criterion for the grouping of damage scenarios is the resulting adverse effect on the road user, especially considering the same or similar chain of effects running from threat scenario to damage scenario. For example, if there are two damage scenarios, *driver's name disclosed to unauthorized parties* and *driver's address disclosed to unauthorized parties*, they can be combined into one damage scenario *personal data stolen*. The specific details are not lost in the TARA, since the respective threat scenarios will contain the *driver's name* and the *driver's address*.
- *Are enough threat scenarios*[2] *identified?* The threat identification step should be completed as thoroughly as possible, as it is the essence of TARA. However, once the threat catalogs have been reviewed and a good deal of effort has been invested in brainstorming sessions, it may be difficult to come up with more threats. In addition, if the threat scenarios already identified present cybersecurity risks that need to be mitigated, it may be sufficient to leave those threats in the TARA and focus on applying controls to the item. As a living document, the TARA can be updated at any time, for example, if further technical discussions identify new threats or new vulnerabilities. Even after the controls have been applied, the TARA may be reevaluated.
- *Is the technical level of the attack paths sufficient?* The attack paths must be accurate enough for the cybersecurity team to make an evaluation of attack feasibility as accurate as possible. If the technical level achieved is sufficient for an accurate attack feasibility evaluation, then it may be reasonable not to go deeper into the technical level. For example, if the step *do fuzzing on UDS from DID 0x0 to DID 0xFFFF on CAN using SID 0x2E* of an attack path is sufficient to perform an evaluation, the intermediate steps such as *connect a CAN debugger* or *develop a fuzzing tool* can be omitted.

4.3.3 "The 300-Days TARA"—The Inefficient or Overblown TARA

A large TARA may be necessary when working at a very detailed level. For example, when analyzing hundreds of individual signals at several ECUs or for the entire

[1] See definition of *damage scenario* in Sect. 6.4.

[2] See definition of *threat scenario* in Chap. 7.

vehicle. It can also happen that the overall high effort is related to the TARA process as a whole, but some parts might get much higher effort than others. For instance, consider a scenario where the cybersecurity team is tasked with providing traceability of each message or asset to the corresponding system requirements, employing a specialized tool. In such cases, the effort invested in achieving traceability may outweigh other aspects of the TARA process. Based on our experience, we argue that an overly exhaustive TARA is not essential in the majority of cases. Simply increasing the level of detail achieved at a certain point does not necessarily improve the overall outcome of a TARA because an overly complex structure induces more problems than a smaller, clearer TARA can solve. It is important to maintain a balance and focus on those aspects that have the most significant impact on the outcome. This pragmatic approach ensures that resources are allocated efficiently, avoiding unnecessary efforts that do not contribute substantially to the effectiveness of the TARA.

In essence, the goal is to strike a balance between thoroughness and practicality. While it is important to be comprehensive in identifying and assessing threats, it is equally important to ensure that the process is manageable and aligned with the project's resources and timeline. Overly extensive TARAs can lead to *analysis paralysis*, where the sheer volume of information and complexity of the process hinder progress and decision-making. Moreover, it is crucial to remember that TARA is not a one-time activity, but a continuous process that should be revisited throughout the life cycle of the product. This allows for adjustments and refinements as the product, its environment, and the threat landscape evolve. Therefore, it is more beneficial to have a manageable and repeatable TARA process that can be regularly updated, rather than an overly extensive one-time analysis.

4.4 Overview of the TARA Process

This book serves as a comprehensive and practical guide to automotive TARA, drawing from our extensive professional experience in the field and adhering to the ISO/SAE 21434 standard. While maintaining strict adherence to the standard, we go deeper into specific steps of the TARA process, breaking down the framework outlined in ISO/SAE 21434.

The formal threat analysis and risk assessment steps (also called methods) and work products according to ISO/SAE 21434 are:

- Asset identification
  - [WP-15-01] Damage scenarios
  - [WP-15-02] Assets with cybersecurity properties
- Threat scenario identification
  - [WP-15-03] Threat scenarios

- Impact rating
  - [WP-15-04] Impact ratings with associated impact categories
- Attack path analysis
  - [WP-15-05] Attack paths
- Attack feasibility rating
  - [WP-15-06] Attack feasibility ratings
- Risk value determination
  - [WP-15-07] Risk values
- Risk treatment decision
  - [WP-15-08] Risk treatment decisions

In Table 4.1 we summarize the TARA steps as described in the ISO/SAE 21434 standard, including the prerequisites, further supporting information, and work products, according to the standard.

We approach threat analysis and risk assessment in a more general way, but still address each work product of the ISO/SAE 21434 standard. We group the TARA methods of the ISO/SAE 21434 into five overall steps: *Scope*, *Assets*, *Threat Scenarios*, *Risk Evaluation*, and *Risk Treatment Decision*. The following chapters of this book tackle each of these five steps. The step scope covers the information needed to start a hands-on TARA project, just as understanding the item definition, describing the assumptions, and, if necessary, creating an attacker model. The step scope is not part of the ISO/SAE 21434, at least not in the way we meant in this book because the item definition, including assumptions, is a prerequisite to the chapters starting with TARA at ISO/SAE 21434. The generic process and the mapping to the work products of the standard are shown in Table 4.2.

The overall TARA process followed in this book is depicted in Fig. 4.2.

Order of the TARA Steps

ISO/SAE 21434 defines TARA as a modular process where each method of TARA can be performed in any order. The order depicted in Fig. 4.2 follows a natural TARA process, where each step builds the necessary information used in the next step. However, in certain projects it may be perfectly valid to start at any other step. For example, starting with the definition of damage scenarios and impact rating, followed by the identification of threat scenarios, and then performing the identification of assets.

Table 4.1 ISO/SAE 21434 TARA steps including prerequisites, further supporting information, and resulting work products

TARA step	Prerequisites	Further supporting information	Work products
15.3 Asset identification	Item definition [WP-09-01]	Cybersecurity specifications [WP-10-01]	[WP-15-01] Damage scenarios, [WP-15-02] Assets with cybersecurity properties
15.4 Threat scenario identification	Item definition [WP-09-01]	Cybersecurity specifications [WP-10-01], Damage scenarios [WP-15-01], Assets with cybersecurity properties [WP-15-02]	[WP-15-03] Threat scenarios
15.5 Impact rating	Damage scenarios [WP-15-01]	Item definition [WP-09-01], Assets with cybersecurity properties [WP-15-02]	[WP-15-04] Impact ratings with associated impact categories
15.6 Attack path analysis	Item definition [WP-09-01] or cybersecurity specifications [WP-10-01]	Weaknesses from cybersecurity events [WP-08-04], Threat scenarios [WP-15-03], Weaknesses found during product development [WP-10-05], Architectural design, Previously identified attack paths [WP-15-05], if available, Vulnerability analysis [WP-08-05]	[WP-15-05] Attack paths
15.7 Attack feasibility rating	Attack paths [WP-15-05]	Architectural design, Vulnerability analysis [WP-08-05]	[WP-15-06] Attack feasibility ratings
15.8 Risk value determination	Threat scenarios [WP-15-03], Impact ratings with associated impact categories [WP-15-04], Attack feasibility ratings [WP-15-06]	None	[WP-15-07] Risk values
15.9 Risk treatment decision	Item definition [WP-09-01], Threat scenarios [WP-15-03], Risk values [WP-15-07]	Cybersecurity specifications [WP-10-01], Previous risk treatment decisions of the item or component, or of similar items or components, Impact ratings with associated impact categories [WP-15-04], Attack paths [WP-15-05], Attack feasibility ratings [WP-15-06]	[WP-15-08] Risk treatment decisions

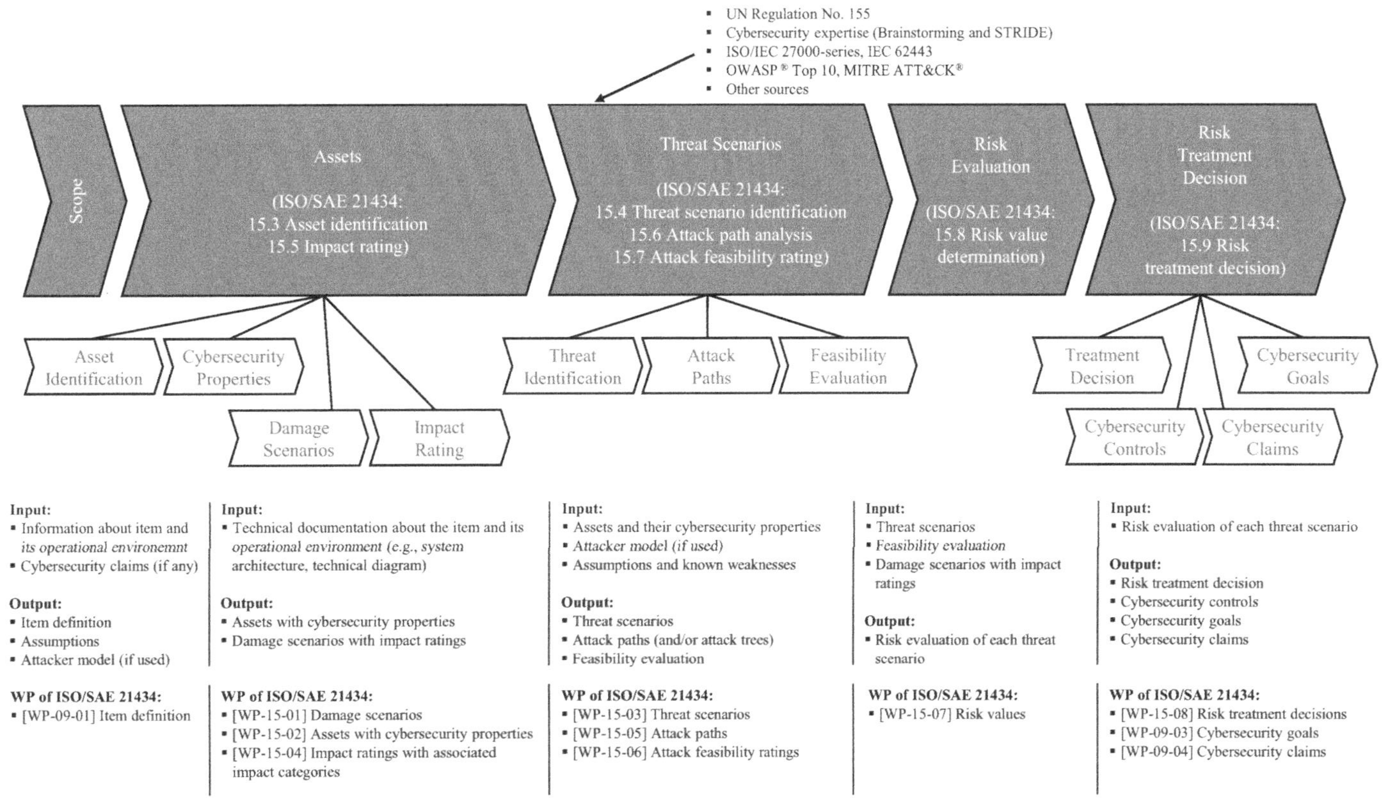

Fig. 4.2 Overall threat analysis and risk assessment process

Table 4.2 Generic TARA process and mapping to ISO/SAE 21434 work products

Generic TARA process	Satisfied ISO/SAE 21434 Work Product
Scope of the TARA, including: • Item definition • Operational environment • Assumptions • Attacker model	[WP-09-01] Item definition
Assets, including the steps: • Asset identification • Cybersecurity properties • Damage scenarios • Impact rating	[WP-15-01] Damage scenarios [WP-15-02] Assets with cybersecurity properties [WP-15-04] Impact ratings with associated impact categories
Threat scenarios, including the steps: • Threat identification • Attack paths • Feasibility evaluation	[WP-15-03] Threat scenarios [WP-15-05] Attack paths [WP-15-06] Attack feasibility ratings
Risk evaluation	[WP-15-07] Risk values
Risk treatment decision, including the steps: • Treatment decision • Cybersecurity controls • Cybersecurity goals • Cybersecurity claims	[WP-15-08] Risk treatment decisions [WP-09-03] Cybersecurity goals [WP-09-04] Cybersecurity claims

4.5 Tooling

Each organization has the freedom to choose the tool that best meets its specific needs. A risk management tool can significantly optimize the overall process by providing a standardized methodology across the organization. For example, catalogs of assets, threats, and countermeasures can be stored in the tool and shared across the organization, promoting a consistent approach to risk management. In addition, when the TARA process is properly integrated with the development process, the TARA tool can provide traceability to the requirements and specifications for the system being analyzed. This integration can streamline the process, making it easier to track and manage risk throughout the product life cycle.

We have worked with several commercial tools explicitly designed for TARA for OT systems and the automotive industry, as well as spreadsheet tools developed internally. While we do not recommend any particular commercial tool in this book, it is worth noting that commercial tools can be very practical and can significantly improve the overall efficiency of the TARA process. However, these tools require an initial effort to understand their functionality and, most importantly, to configure them to meet the specific needs of the organization. The tool should be tailored to the exact needs of the cybersecurity team and the company's policies. On the other hand, spreadsheet tools can be very beneficial because of their inherent flexibility. They allow for customization and adaptation to specific needs. They are also widely used and understood, reducing the learning curve for team members. And they can be easily shared and updated, making them a practical choice for many organizations.

It is important to remember that tool selection is only one aspect of the TARA process. Regardless of the tool used, the key to effective risk management is a well-defined process, a clear understanding of the organization's risk tolerance, and a commitment to continuous monitoring and improvement. The tool should support these objectives, not dictate them.

Tip
For those readers who do not have a TARA tool, we provide a TARA spreadsheet template at this link: https://doi.org/10.1007/978-3-662-69614-9_4.

References

1. Puder, A., Henle, J., Sax, E.: Threat assessment and risk analysis (TARA) for interoperable medical devices in the operating room inspired by the automotive industry. Healthcare (Basel) **11**(6) (2023). https://www.ncbi.nlm.nih.gov/pmc/articles/PMC10048460/
2. EVITA Project: E-safety vehicle intrusion protected applications (EVITA). https://www.evita-project.org/
3. Lautenbach, A., Mafijul Islam, C.: HEAling vulnerabilities to ENhance software security and safety (HEAVENS)—deliverable d2—security models (2016). https://autosec.se/wp-content/uploads/2018/03/HEAVENS_D2_v2.0.pdf
4. ETSI: ETSI TS 102 165-1 V5.2.5—CYBER; Methods and protocols; Part 1: Method and pro forma for Threat, Vulnerability, Risk Analysis (TVRA). Standard (2022). https://www.etsi.org/deliver/etsi_ts/102100_102199/10216501/05.02.05_60/ts_10216501v050205p.pdf
5. VerSprite: PASTA Threat Modeling: The Process for Attack Simulation and Threat Analysis (eBook) (2022). https://versprite.com/cybersecurity-listings/offsec/pasta-threat-modeling/
6. ISO/IEC 27001:2022 Information security, cybersecurity and privacy protection—Information security management systems—Requirements. Standard, International Organization for Standardization/International Electrotechnical Commission (2022). https://www.iso.org/standard/82875.html
7. ISO/SAE 21434:2021 Road vehicles—Cybersecurity engineering. Standard, International Organization for Standardization/Society of Automotive Engineers, Geneva, CH (2021). https://www.iso.org/standard/70918.html

Chapter 5
Defining the Scope of a TARA Project

Abstract With the scope definition, we set the foundation of a TARA project. The scope defines how the TARA will be addressed. In addition to describing the item definition and its operational environment, at this step we describe technical assumptions for the TARA, e.g., about existing controls, existing vulnerabilities, cybersecurity goals, or cybersecurity claims. Any special features of the TARA project are described in the scope. In addition, we describe the attacker model and its capabilities. We also define the impact categories for the damage scenarios in the context of the scope. In this chapter, we also introduce the practical case studies that will be presented throughout the book.

This chapter is about the very first step of a TARA. We define the scope of a TARA project, i.e., the boundaries of the system to be analyzed, including all essential information for performing TARA, such as assumptions, details from existing projects or known vulnerabilities, together with the definition of the attacker model. A core element of the scope of a TARA is the description of the system to be analyzed. The description of the system to be analyzed corresponds to the *Item Definition*, or component specification, according to the ISO/SAE 21434. In other risk management frameworks, the system to be analyzed is usually referred to as the *Target of Evaluation (ToE)* or *System Under Consideration (SUC)*. Defining the scope is a critical task that should not be underestimated, as it lays the foundation for the entire TARA process. Establishing a clear scope can save time and effort later in the project by providing a clear direction for the TARA analysis. In our experience with numerous TARA projects, we have seen common mistakes resulting from poorly defined scopes, leading to the need for major changes to fundamental aspects at later stages. The scope in the overall TARA process is depicted in Fig. 5.1.

In this chapter, we address the key aspects of scope definition, emphasizing the importance of a clear description and understanding of the operational environment in which the system to be analyzed operates. We explain the importance of assumptions and highlight the benefits of explicitly defining out-of-scope elements, which can greatly help to optimize the subsequent analysis. We also approach scope definition from a practical perspective, exploring typical scenarios for different types of systems. We examine scope considerations when dealing with embedded devices as well as non-embedded systems, such as IT infrastructure and manufacturing environ-

R. do Carmo and A. Schlensog, *Automotive Threat Analysis and Risk Assessment in Practice*, https://doi.org/10.1007/978-3-662-69614-9_5

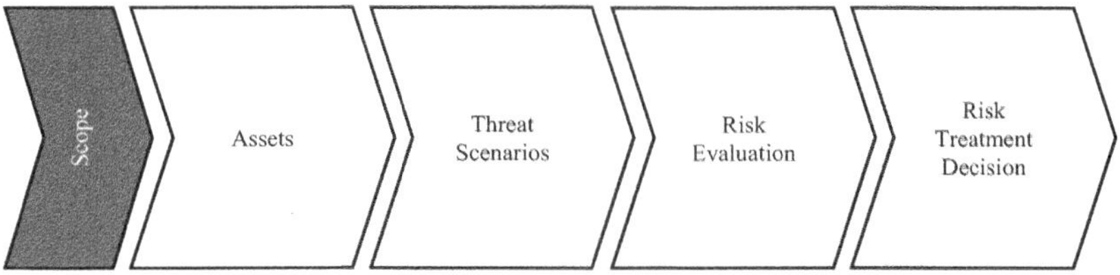

Fig. 5.1 The scope step in the overall TARA process

ments. By providing practical insights and guidance, we aim to equip readers with the knowledge necessary to effectively define scope for a wide range of TARA projects. As a practical demonstration of the concepts discussed, this chapter also introduces the practical case studies that will be presented throughout the book. These case studies serve as valuable illustrations that apply the principles and techniques covered in each chapter to real-world scenarios, thereby enhancing the reader's understanding and ability to effectively apply the TARA process.

Item Definition According to ISO/SAE 21434 In the context of ISO/SAE 21434, an *item* is a system that implements a function at the vehicle level and is composed of *components*. TARA can be performed at both the item level and the component level. It is important to note that the explicit identification of the item or component is not described as part of the TARA process outlined in *Chapter 15* of the standard. Instead, it serves as a prerequisite for the TARA process itself. In the case of an item, its definition is addressed as a separate step in the standard called *Item Definition* and referred to as the work product *[WP-09-01] Item Definition*. A component is defined with cybersecurity specifications. In this work, we take a broader approach by including item definition as an integral part of the TARA process. By doing so, we aim to present a more comprehensive and adaptable TARA process that can be applied to different systems. It is worth noting that many ISO/SAE 21434 TARA projects do not consistently emphasize certain important aspects, such as a clear description of the operational environment in which the item operates. In addition, critical assumptions, such as whether to include cybersecurity controls as a premise, need to be thoroughly discussed as part of the scope. Recognizing these gaps, we aim to address these aspects within the scope of the TARA process to ensure a comprehensive analysis.

Performing TARA at item level or component level?

A TARA can be performed at *item level* and at *component level* according to ISO/SAE 21434. The process of performing a TARA for an item and for a component is almost identical, except for minimal differences. For this reason, we do not treat the item and the component as separate systems in this book. Although there may be minimal differences between the two TARA approaches, such as the lack of an explicit *item definition* for components, we make sure to explicitly mention these differences when necessary. Otherwise, we use the term *item* in a general sense throughout the book, including both items and components.

Necessary input for this chapter:

- Information about the item and its operational environment
- Cybersecurity claims (if any)

The outcome of this chapter is the scope of TARA, including:

- Item definition
- Assumptions
- Attacker model (if used)

5.1 How to Define the Scope

When starting a TARA project, the overall target system or *item definition* may be clear from the beginning of the project. However, it is possible that the item definition is not available or is not described clearly enough. To perform TARA as efficiently as possible, all functions, and components of the overall system must be carefully analyzed and described. Parts external to the item, i.e., related to the operational environment, that may be relevant in the context of cybersecurity should also be analyzed. According to requirements *[RQ-09-01]* and *[RQ-09-02]* of ISO/SAE 21434, the item definition must consist of:

- *Item boundary:* The item shall be clearly distinguished from its operational environment, including interfaces with other items internal to the vehicle and with E/E systems external to the vehicle.
- *Item functions:* Description of the intended behavior of the item during the life cycle phases, such as product development and testing, production, and post-production, and the vehicle functionality realized by the item.
- *Preliminary architecture:* A description that can include internal components of the item and their connections, and all external interfaces of the item.
- *Cybersecurity-relevant information:* Description of any information about the operational environment relevant to cybersecurity, such as interaction of the operational environment with the item that can help by the threat modeling step. Assumptions about cybersecurity can also be included.

We recommend starting with an overall diagram of the item to be analyzed by the TARA, showing its interaction with the rest of the environment. For embedded devices, such as an ECU, this may be an E/E diagram showing the high-level external interfaces of the item, high-level information about key functional blocks, or software components. We also recommend that the cybersecurity analyst or cybersecurity team performing the TARA obtain enough technical documentation at this point to have general knowledge of the item being analyzed. The cybersecurity team and the field experts involved with the item (e.g., system architects, software experts, project managers, and product cybersecurity engineers) need to identify the boundaries of the item. We have identified a group of elements that, in our experience, should be part of any TARA scope definition: item and in scope, operational environment, out-of-scope systems or components, impact categories, attacker model, and assumptions. We describe them in detail as follows.

Item and In Scope The item definition, especially the clear description of its boundaries and functions, must be available for the execution of TARA. For the sake of documentation and to provide a better understanding to the whole team working on the TARA project, we find it beneficial to explicitly mention elements, e.g., details on functionalities, details on E/E systems, details on external systems, that have to be in the scope of TARA.

Operational Environment A system to be analyzed for a TARA, such as a brake ECU, is typically not deployed as a standalone device, disconnected from the rest of the vehicle. This may be true for very simple devices, such as small sensors, but usually the analyzed item interacts with many other devices connected to it, in other words, with its operational environment. For this reason, it is of utmost importance to describe the operational environment and keep its operational functions present at the moment of performing TARA. For example, a brake ECU can be designed to operate securely by protecting the main CAN signals to the rest of the vehicle from tampering. In the context of autonomous driving, it might be possible, from a cybersecurity perspective, to send valid, correctly signed signals to the braking ECU by compromising the autonomous driving ECU. In this simplified example, the signals from an autonomous driving ECU may be within the scope of the TARA of the braking ECU, even if the autonomous driving ECU itself is not within the scope. As described in ISO/SAE 21434, the description of the operational environment and its interactions with the item can influence the identification and analysis of relevant threat scenarios and attack paths [1]. For example, in the brake ECU example, consideration of the threat of impersonation of an autonomous driving ECU may lead to a recommendation to implement mutual authentication of ECUs as a cybersecurity control at the end of the TARA. Suppose there is a new project for the same brake ECU platform, and the brake ECU is not used in an autonomous driving vehicle. In the case of brake ECU reuse analysis, the operational environment has changed significantly, and it needs to be carefully analyzed. The mutual authentication control may no longer be necessary in this new context of the brake ECU. In addition, the new operational environment may introduce different features that were not present in the previous project (e.g., new debugging capabilities).

Out of Scope It is important to explicitly describe parts of the analyzed item, functions, or elements related to the operational environment that, for various reasons, are outside the scope of the TARA project. For example, hardware debugging interfaces such as JTAG may be present at an early stage of the item's development. However, if it is ensured that these interfaces will be removed for series production, they can be left out of the scope of the TARA. A TARA, as a living document, is likely to be reviewed throughout the life cycle of the item, and therefore it is important to describe the out-of-scope components to avoid future questions as to why they were not included. This may be especially true in the case of reuse, where an existing TARA of the item to be reused is used as the baseline for a new analysis. In addition, in the case of an assessment or audit, a well-documented description of out-of-scope features can prevent unpleasant situations. We emphasize that good documentation is an essential requirement of a TARA.

Impact Categories The impact categories are applied to the damage scenarios to assess their impact. Even if the identification of the damage scenarios is a step that is done later in the TARA around asset identification, we can still define the impact categories as part of the scope definition. The definition of the impact categories leads to a more profound understanding of the objectives of the TARA and can help in the definition of the scope itself. For example, ISO/SAE 21434 requires that damage scenarios for the impact categories of safety, financial, operational, and privacy be evaluated against potential adverse consequences for road users. However, other categories, even for stakeholders other than road users, are allowed. For example, a TARA for the ECU of a vehicle may be performed from the perspective of the road user to comply with the standard, but also from the perspective of the supplier, affecting categories such as loss of intellectual property, financial loss, or loss of image. Using intellectual property as an example, parts of the data or compiled software in the ECU can be evaluated differently, e.g., enforcing confidentiality as a cybersecurity property, as it would be if the TARA were performed only from the road user perspective. The road user is not affected at all if the supplier's intellectual property is compromised.

Attacker Model Defining the attacker model helps to identify the appropriate threats and evaluate their feasibility later in the TARA process. We discuss this topic in detail in Sect. 5.6 of this chapter.

Assumptions Assumptions of any kind belong in the first step of TARA because they establish the foundation upon which the entire TARA process is built. We discuss this topic in detail in Sect. 5.2 of this chapter.

An overall summary of the scope of a TARA as proposed in this book is shown in Fig. 5.2. We provide an example of elements defined in the scope definition in Table 5.1. When performing a TARA, it is recommended to use Table 5.1 directly as a template for the scope. It is also helpful to have a diagram where it is easy to identify the scope at a glance (e.g., by marking the block of the item and its interfaces with a green color, and out-of-scope elements with a red color).

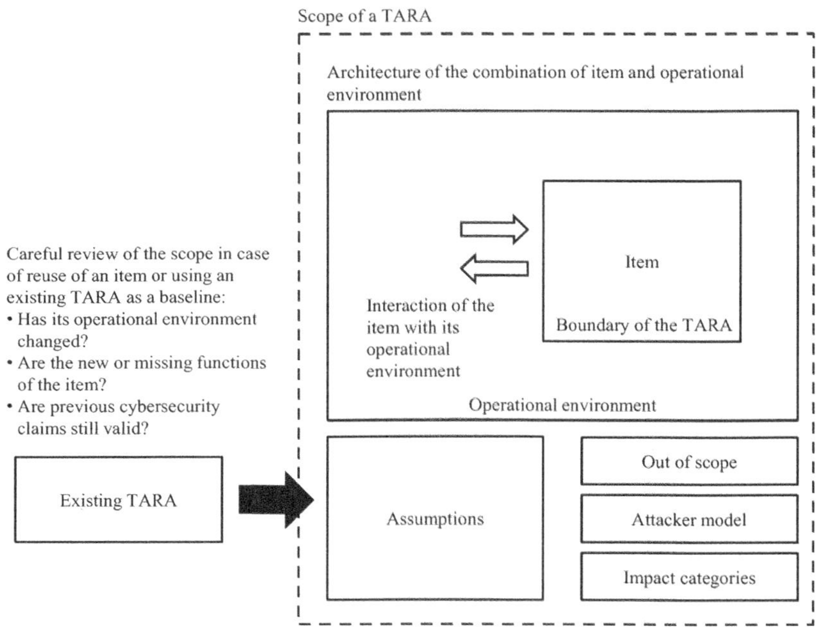

Fig. 5.2 Scope of a TARA

Table 5.1 Example of a table that includes the relevant elements of the scope definition of a TARA

General Description	
Item:	Description of the item (or component in case of a component TARA)
In scope:	Details about part of the item or component that are worth to mention as explicitly in scope
Operational environment:	Description of the operational environment of the item
Out of scope:	Examples could be debugging interfaces being removed in production, or off-the-shelf components interacting with the item in scope
Impact categories:	Mandatory for Road User: Safety, Financial, Operational, Privacy Optional for OEM/Supplier: Intellectual Property, Financial, Image
Attacker Model	
Threat level:	Attackers have a threat level *Severe*
In scope:	If needed, modifications to the generic attacker model can be described here by explicitly increasing the scope
Out of scope:	The attacker model can be tailored for a particular project, e.g., by removing some threat actors
Assumptions	
Description of assumption number 1	
Description of assumption number 2	

5.2 Assumptions

Assumptions are an important element of the scope, not only because they define fundamental aspects of TARA, but also because they improve the overall level of documentation of the TARA project. We argue that, in general, there is no such thing as a bad approach to TARA, except bad documentation or no documentation at all. Assumptions allow stakeholders to align their expectations of the TARA process and its outcomes. By clearly stating assumptions, it is possible to manage stakeholder expectations and avoid misunderstandings or conflicts later on.

Assumptions can be of all kinds. For example, the item may be under development, and the current state of development is that a different communication protocol will be integrated for series production. In this case, it can be assumed that only the final communication protocol will be considered for the TARA. Assumptions can also be made about assets. For example, the ECU to be analyzed may store personally identifiable information (PII) in other projects, but not for the particular project being used for the TARA. Assumptions about the process of the TARA can also be written. For example, the TARA is performed by a third party company that performs all the steps of the TARA, as shown in our process, except for the treatment decision. Another example of a process-level assumption can be the limitation of the cybersecurity properties applied. The TARA can be realized using a template where six cybersecurity properties appear along the document, such as confidentiality, integrity, availability, authenticity, non-repudiation, and accountability. For a particular TARA project, it may be sufficient to consider only three of these, e.g., confidentiality, integrity, and availability, and leave the other security properties blank.

Cybersecurity Claims An important element of the assumptions are the cybersecurity claims. Cybersecurity claims are defined in ISO/SAE 21434 as statements about risk, and they are specified when the risk treatment decision about a high risk is to accept (or retain) or share the risk. If there is a pre-existing TARA or any type of risk assessment with cybersecurity claims, they need to be considered for the current TARA. An important side effect of considering cybersecurity claims is to avoid analyzing combinations of assets-threat scenarios where cybersecurity claims are already being applied.

5.2.1 Important Assumption: Shall Existing or Planned Controls for the Item Be Considered for the TARA?

There are two basic approaches to the cybersecurity controls of an item, usually related to the stage of development when the TARA is performed. One is the TARA performed at the beginning of a product's development process. At this point, the overall cybersecurity goals may be defined, but there are often few or no concrete, technical cybersecurity controls planned. On the other hand, there is the TARA, which is conducted at an advanced stage of product development, and at least some cybersecurity controls are already built into the system design. We emphasize the importance of this assumption because we observe a discussion about whether to include controls in almost every TARA project.

No Cybersecurity Controls Planned In the early stages of product development, there are typically few or no concrete cybersecurity controls planned. In fact, one of the primary reasons for conducting a TARA is to define the cybersecurity requirements, architecture, controls, and overall needs based on the identified risks. For this reason, we recommend assuming early on that no technical cybersecurity controls are part of the system. This leads to the identification of as many threats as possible. This approach is also valid for a TARA that is performed at an advanced stage, but for backward documentation purposes (for example, because no TARA was performed before).

Planned Controls If they are cybersecurity controls already planned for the item or component, they can be considered when performing the TARA. The advantage of considering controls is that many threats are already mitigated by the controls, so the size of these TARAs will tend to be smaller than those without controls. However, we recommend considering all possible threats when using this approach to minimize the risk of overlooking threats. It may also be useful for documentation purposes, for example when dealing with regulatory assessments such as UN Regulation No. 155, to explicitly mention the threats covered by that regulation's catalog.

5.3 Scope for Automotive Embedded Devices

The electrical and electronic (E/E) architecture of a vehicle defines the electrical supply of the different ECUs in the vehicle, as well as their electronic integration with each other. A typical E/E architecture of a vehicle describes how and over which communication buses the ECUs are interconnected. The ECUs are typically separated in domains, for example, in a body controller domain, a powertrain domain, and an infotainment domain. The most common communication protocol used in vehicles for several decades is the Controller Area Network (CAN), today in its modern variant CAN Flexible Data-Rate (CAN FD). Local Interconnect Network (LIN) and FlexRay are common communication protocols as well. For high-data rates uses, like for infotainment systems or advanced/autonomous driving systems, MOST, and Ethernet are commonly protocols as well. For the analysis of TARA, two different approaches are typically considered: a) TARA for standalone ECUs, b) TARA at vehicle level. ECUs are produced by suppliers, and they are usually developed as a standalone product, at least for the same OEM or vehicle platform. This means that vehicle platforms of the same OEM typically share many ECUs in common, but also with vehicles from different OEMs. The supplier has in this case interest in performing a TARA, either for the ECU as platform-independent baseline, or for the ECU applied for a particular project. On the other hand, the OEM, who has the overall responsibility for the vehicle, is also interested in analyzing the risks for the complete architecture.

5.3.1 *TARA for Standalone ECUs or Functional Systems*

The typical scope for a single ECU, or a combination of ECUs that form a closed functional system, is to analyze the communication interfaces to the ECU and its internal architecture. Communication interfaces can be used as attack vectors, either for threats originating from the in-vehicle network or from networks outside the vehicle. Internal components of the ECU are also considered attack vectors, or at least vulnerable elements, such as an unprotected memory block or debugging connector. For simple ECUs, communication interfaces may be limited to a single CAN bus and on-board debugging interfaces such as JTAG. Examples of simple ECUs include door or sunroof controls, seat heater controls, rain sensors, or ECUs controlling smaller sensors such as temperature and pressure. In more complex ECUs, such as a telematics unit, the communication interfaces may include multiple CAN, CAN-FD, and FlexRay interfaces to the vehicle, inter-controller connections, and external interfaces to Bluetooth, Wi-Fi, USB, or in-car technologies such as Apple CarPlay or Android Auto. While the focus of the scope is typically on the physical boundaries of the item, it is valid to include in the scope details about the upper layers and services that communicate over the interfaces. This information is useful for analyzing the asset. Examples of typical upper layer services can be the popular Unified Diagnostic Services (UDS) running over CAN, FlexRay, LIN or MOST buses, proprietary diagnostic protocols, or even modern approaches such as the Scalable Service-Oriented Middleware over IP (SOME/IP), which enables a service-oriented architecture similar to Internet Web services.

5.3.1.1 Scope of a *Generic TARA* and Distributed Responsibility Between OEM and Suppliers

Particularly for single ECUs or systems supplied by suppliers, there is often discussion of sharing the responsibility for performing a TARA between the OEM and the supplier (or between suppliers). Conducting a TARA is an arduous task, requiring deep technical knowledge of the item and its operating environment. Suppliers may consider it convenient to perform a *generic TARA* that can be shared, for example, among several projects for the same product type or product family. This kind of generic TARA certainly does not meet the expectations of a TARA according to ISO/SAE 21434 because it cannot consider the operational environment and specific technical details of a project. However, a generic TARA can be created that makes assumptions about the operational environment. The OEM can later take the generic TARA from the supplier and complete the TARA with the details of the operational environment and system in which the item will be used. In other words, the OEM can perform gap avoidance and bridge the gap between the generic TARA that considers the item in a fictitious operational environment and the item as applied to the OEM's current project.

The German Association of the Automotive Industry (VDA) proposes a recommendation for a cybersecurity interface agreement (CSIA) [2], where there is a proposal for different types of TARA between the OEM and the supplier (Tier 1). The OEM is fully responsible for performing the TARA (i.e., TARA according to ISO/SAE 21434), while the supplier only supports the activity. However, the interface agreement proposes the creation of an out-of-context or component TARA provided by the supplier (Tier 1) for the delivered component, software, or hardware to verify the OEM's cybersecurity goals and requirements.

5.3.2 TARA at the Vehicle Level (E/E Architecture)

The scope of a TARA at the vehicle level (also called *system-level TARA* or *E/E architecture-level TARA*) has a similar principle as for a single device, i.e., all possible attack vectors are of interest. However, there are two important aspects to defining the boundaries of the system, i.e., the item. First, threats originating from unprotected internal components (e.g., internal ECUs) are typically in scope, but TARA is limited to communication with these components and does not look deeper into them. The internal components act as black boxes. Second, communication with the outside world is usually in scope because it is probably the most interesting aspect of such a TARA. However, the scope is limited to the communication data, i.e., the protocols that talk to the E/E architecture, and not to attack vectors originating from compromised back-end servers or cell phones. It should be noted that malicious messages present in the vehicle due to a manipulated internal component or a compromised back-end server that compromises an asset of the vehicle-level TARA must be within the scope of the TARA. However, the vehicle-level TARA does not differentiate between, for example, a back-end server being compromised by a *bad USB* or someone gaining root access to a back-end server. The goal of the vehicle-level TARA is to provide cybersecurity goals for these external systems to mitigate identified vehicle-level risks.

Typical examples of elements in the scope of a modern E/E architecture can include:

- Complete in-vehicle E/E architecture communicating all vehicle domains
- In-vehicle communication architecture, protocols, and services
- Sensors and actuators
- Cell phone applications (OEM or third party)
- Cell phone platforms such as Apple CarPlay or Android Auto
- OEM, suppliers or third-party remote services
- Location services (GNSS)
- Cellular, Wi-Fi and Bluetooth communication

- eCall and concierge services
- Microphones
- Over-the-air software (SW) updates and third-party applications
- Keyless entry/Key fob
- Workshop diagnostics tools/Debugging tools
- OBD-II

5.4 Scope for IT/OT Systems for Vehicle-Backend Architectures

A modern connected car can be compared to a cell phone on wheels [3]. In fact, a modern connected car head unit has at least the same capabilities as a modern cell phone, such as one or more eSIM interfaces for cellular communication, Wi-Fi capabilities, powerful computing capabilities capable of running Linux operating systems and processing large amounts of data, for example, for using real-time navigation services. A modern vehicle can connect directly to cell phones via Bluetooth or Wi-Fi, and even to remote services for cell phone applications. For example, an application can request the vehicle's location, lock or unlock the doors, provide technical information such as battery charge status, and even send addresses to the navigation system. In the near future, even the so-called vehicle-to-everything (V2X) communication capabilities will be part of the connected car. V2X technology allows the car to interact with other vehicles, for example, to exchange vehicle position and speed in order to avoid collisions. All of these highly connected features rely on a complex ecosystem of actors that we need to carefully analyze when performing a TARA at the vehicle backend level.

What do we want to achieve with a system-level/vehicle-level TARA?
The main reason for implementing TARA at the system level, i.e., at the vehicle-backend communication level, is to understand the overall risks of the whole system and to derive *cybersecurity goals* for the elements that comprise the system. At the architectural level, the TARA element is defined by the communicating elements as closed systems. For example, the communication channel from a vehicle to a cell phone and the data exchanged may be distinct assets belonging to the two elements (cell phone and vehicle). In a TARA at this level, for example, we do not look at the electronics of the cell phone. If a resulting cybersecurity goal of the TARA is to protect personal data received by the cell phone from the vehicle from unauthorized access as it travels to the application and the backend server, it may be within the scope of a specific TARA for the cell phone to analyze the internal cybersecurity risks of the cell phone to achieve this goal. Looking at the connected car, we can still consider

some of the most relevant internal components, i.e., the most relevant ECUs such as the head unit, as part of the components as black boxes for the TARA.

5.5 Scope for Manufacturing Plants

In this book, we analyze the supply chain as part of an IT/OT system, applying the same principles as for a TARA for vehicles. Industry 4.0 is transforming the classic, unconnected production plants into a highly automated and interconnected ecosystem. Many of today's and tomorrow's production plants are connected and rely on cloud services such as Amazon® Web Services or Microsoft® Azure. Advanced technologies such as artificial intelligence (AI) in manufacturing are no longer science fiction; in fact, the market for AI in manufacturing is expected to grow to USD 16.3 billion in the next few years [4]. On the other hand, we want to address the TARA process in a generic, holistic approach to comply with UN Regulation No. 155 and ISO/SAE 21434. UN Regulation No. 155 requires OEMs to demonstrate the capabilities of their CSMS in all phases of the product life cycle, namely the development phase, the production phase, and the post-production phase. The ISO/SAE 21434 standard, which aims to achieve a CSMS for compliance with UN Regulation No. 155, requires in its requirement *[RQ-12-02]* that a production control plan be established to ensure that post-development cybersecurity requirements are applied to the item or component and that vulnerabilities cannot be introduced during production. The control plan for production includes the steps necessary to apply the post-development cybersecurity requirements, the production tools and equipment used, cybersecurity controls to prevent unauthorized changes during production, and validation methods to confirm that the post-development cybersecurity requirements are met [1]. Traditional approaches to risk management in production facilities typically follow the IEC 62443 international standard for OT systems and the well-known ISO/IEC 27000-series of international standards for information security.

5.6 Defining the Attacker Model

There are a number of good reasons for having an attacker model in TARA. One of the key points of modeling the attacker is to help the entire team, including management and technical experts, better understand the overall goals of the TARA project. Moreover, defining the attacker allows for targeted threat limitation based on specific characteristics. For example, for complex systems with items composed of multiple devices, complex microcontrollers that handle many types of sensitive assets, and strong cybersecurity goals such as protecting personal information, the concept of limiting the threats may not be necessary at all. For such projects, we probably want to consider as many threats as possible, regardless of the threat actor. On the other hand, there are systems where either the complexity of the device is rather small. For example, a small device that controls only a few non-safety signals, with only a single CAN interface that is not part of the safety CAN, and where the only cybersecurity goal is to protect the signals from tampering, it may be useful to model the attacker in such a way that, for example, confidentiality threats are not considered.

In addition, for documentation purposes for future analysis (e.g., reuse), it is important for the cybersecurity team to understand how the threats were selected and to get a general sense of the cybersecurity mindset behind the TARA. For example, if there is no attacker model documentation at all and the TARA misses all threats from certain threat actors, a reviewer may question whether the TARA was done correctly or if the threats were simply forgotten.

Attacker model is not part of ISO/SAE 21434

ISO/SAE 21434 has no specific requirements for the attacker model. It is implicitly modeled by identifying threat scenarios where different methods may be considered. It is perfectly correct to skip this section and not include the attacker model in a TARA project. However, defining the attacker has some advantages, for example, it is natural for the E/E and software experts supporting the TARA process to understand the threats when thinking about human actors such as script kiddies, tuners, thieves, or terrorists. This is especially useful when discussing TARA with management teams. Note that we do not follow an attacker-centric model. We limit ourselves to defining an attacker model and assigning capabilities to it that will later help us with threat identification (cf. chapter *Focusing on Attackers* p. 40, in [5]). We follow the attacker modeling approach described in ETSI TVRA [6] in this book, which is also a mentioned threat modeling approach in ISO/SAE 21434. For attacker modeling, we make the necessary adaptations to apply it to ISO/SAE 21434 and to use it later in the TARA threat selection process.

5.6.1 Who is the Attacker?

Before entering into the formal analysis of the threat actor for the purposes of TARA, we describe who the attackers of a connected vehicle or IT/OT infrastructure actually are. We define the following typical adversaries and use them for our attacker modeling approach.

Script Kiddie The term script kiddie is used in the cybersecurity community to describe a non-professional attacker who downloads some software from the Internet and just runs it to see what happens. The attacker is like a child who just wants to test things out without understanding what he is doing. The attacker has no deep technical knowledge, the tools used are usually well known and easy to find on the Internet [7]. The main motivation is curiosity or a small financial gain (e.g., to activate a protected feature in the car).

Tuner Probably one of the most well-known attackers in the automotive world since the existence of the car is the tuner. The hot rods, for example, are tuned cars from the early days. Tuners can be hobbyists who try to improve the appearance and performance of their cars, or professional teams who do the modifications for money. Since the increase in electronics and software in vehicles, the tuner has changed from a technician-tuner to a hacker. The tuner does not necessarily have in-depth knowledge of computer science. He or she may understand how vehicle communication protocols work, and most likely knows the basics of diagnostics via UDS, but he or she still relies on tools found on the Internet or on the black market.

Data Collector Data in general has become one of the most important and valuable assets in today's interconnected world. In fact, 40% of reported cybersecurity incidents recently have resulted in data and privacy breaches [8]. From intellectual property, such as part of the source code of ECU software, to driver privacy data, all kinds of data can be targets for data theft. It is therefore difficult to generalize about a data collector, as the data stolen and the motivation are quite diverse. Criminals looking to make a quick buck may be interested in obtaining private data, such as the routes drivers have taken in recent years, to post on the Internet and blackmail drivers. More professional hackers may be interested in the financial activities of companies, stealing sales data, information about customers, shipping data, or similar information [3]. Therefore, we decided to define the threat actor as an opportunistic hacker who only seeks financial gain by selling data on the illegal market, such as those found on the darknet.

Thief Car theft and break-ins are as old as cars themselves. With modern technology to unlock doors and start engines, such as remote keyless entry systems, car theft and break-ins have also become a cybersecurity issue. According to [8], they are the second cause of all publicly reported automotive-related cybersecurity incidents, accounting for 27.9% of recent incidents.

Saboteur Hacktivists are one of the trendiest cybersecurity threat actors in the IT world observed in recent years [9]. Hacktivists want to draw attention to a particular cause, usually political. They may have quite talented experts working to persuade. Another example of sabotage is competing companies who may fund research teams and direct them to white-hack or pen-test a competitor's vehicle or ECU in order to damage their reputation.

Motivated White-Hat Hacker While this attacker has no malicious intent, the motivated white-hat hacker can cause significant reputational damage to vehicle manufacturers or suppliers. They are highly motivated, such as university researchers who want to publish a paper at a major cybersecurity conference, such as DEFCON or BlackHat, or IT security professionals who want to increase their visibility in the cybersecurity market.

Employee or Ex-Employee The insider is a major threat actor within IT cybersecurity and, of course, for the automotive world. The insider attacker is someone who is against a particular organization and has access to highly confidential information, such as design details of an ECU or even account credentials for critical services. The insider may be motivated by money, ideology, revenge, or coercion [3].

Hacker-for-Hire Actor The Hacking-as-a-Service is another trending hacking modality witnessed lately. Hackers-for-hire are companies that provide offensive cybersecurity capabilities for various reasons, such as vulnerability research and exploitation, malware payload development, technical command and control, operations management, or training and support [9].

Terrorist/Cybercriminal While the term "cybercriminal" may apply to a variety of threat actors, such as the tuner, we focus in this category on highly skilled hacking teams, most often established and controlled by organized criminal gangs. Terrorists or cyberterrorists aim to cause destruction of property, loss of life, and the spread of terror [3].

State-Sponsored Actor These attackers work in highly skilled teams, have enormous resources, and are either directly controlled by, or contracted by, a state. Nations may have various motivations for conducting cyberattacks, such as intelligence and espionage, or war-motivated sabotage of another nation's infrastructure [3]. Advanced persistent threats (APT) are an example of state-sponsored threats.

5.6.2 Attacker Modeling Approach Following ETSI TVRA

ETSI Threat, Vulnerability, and Risk Analysis (TVRA) [6] is a comprehensive, well-known method for threat analysis and risk assessment. It was originally applied to telecommunications systems. Before ISO/SAE 21434, the TVRA method was commonly used in the automotive industry. For example, HEAVENS makes several references and comparisons to TVRA in [10]. We do not follow the full threat modeling methodology of TVRA, in fact, in this chapter, we only deal with the definition of the attacker model. Because of its wide acceptance in the industry, we take advantage of the attacker modeling concept of TVRA. TVRA introduces the concept of *threat agent* for attacker modeling. A threat agent is an entity that can adversely affect assets. Each threat agent has an associated *threat level*. The threat level is an interesting concept introduced by TVRA that combines the ability and motivation of a threat agent to attack an asset.

Table 5.2 contains the motivation levels as described in TVRA. In Table 5.3, we describe the capability levels according to TVRA, and we modify the attack potential mapping to cover the attack potential values required by ISO/SAE 21434. At this point, it is important to remember that the attack potential actually describes the capabilities an attacker must have to execute the attack path for a given asset in the context being analyzed. The mapping in Table 5.3 actually indicates the capabilities that the listed threat actors can achieve. This means, for example, that a script kiddie will not get more knowledge than a layman (if it does, it is no longer a script kiddie), or that it will not get confidential information about the target system. We focus on the threat level value to model our attacker and use the model in the following steps of the TARA process. Table 5.4 describes the final mapping between the motivation and capability of the threat actor according to TVRA. The description of the capability of the threat agent in TVRA is, in our opinion, still difficult to apply practically in the automotive context, since the description of the capabilities is rather subjective and is likely to open discussions during the practical implementation of TARA. We propose in Table 5.5 to fill this gap by mapping the threat level classification from TVRA to different threat actors in the automotive industry. We map the threat actor and the target system to the attack potential that the actor can achieve in order to use the mapping later in the threat selection process. The list of threat actors is not exhaustive or definitive, but it can be used as a reference. Cybersecurity is a very dynamic field because technologies are constantly evolving and because human intelligence is the key driver in the field. In the ENISA Threat Landscape report, we can see how different threat actors emerge and change every year [9].

Table 5.2 Motivation levels according to TVRA

Motivations levels	Description
Very low (indifferent)	The system is considered to have limited to no value to the threat agent; thus the threat agent is very unlikely to attempt any attack on the system
Low (curious)	The system is considered to have minimal value to the threat agent. Threat agents may attempt to attack the system out of curiosity or opportunistic motivation. Non-system deterrents may be sufficient to deter the threat agent from initiating the attack (e.g., due to potential ramifications if the agent can be linked to the attack)
Medium (interested)	The system is considered to have moderate value to the threat agent. The threat agent will attempt to attack the system on a frequent basis. It is also considered unlikely that the threat agent can be deterred from initiating the attack by the existence of non-system deterrents
High (committed)	The system is considered to have significant value to the threat agent. The threat agent will attempt to attack the system on a persistent and frequent basis. It is considered highly unlikely that the threat agent can be deterred from initiating the attack by the existence of non-system deterrents
Very high (focused)	A threat agent has the primary aim of attacking the system

Table 5.3 Capability levels from TVRA with attack potential mapping adapted to ISO/SAE 21434

Capability levels	Description
Very little	Threat agent has almost no capabilities or resources Example mapping to attack potential (Expertise, Knowledge, Equipment): Layman, Public, Standard
Little	Threat agent has very modest capabilities and resources Example mapping to attack potential (Expertise, Knowledge, Equipment): Proficient/Expert, Public, Standard
Limited	Threat agent has modest capabilities and resources Example mapping to attack potential (Expertise, Knowledge, Equipment): Layman/Expert, Restricted, Standard/Specialized
Significant	Threat agent is capable and has significant resources Example mapping to attack potential (Expertise, Knowledge, Equipment): Expert, Restricted/Confidential, Specialized/Bespoke
Formidable	Threat agent is extremely capable and well-resourced Example mapping to attack potential (Expertise, Knowledge, Equipment): Multiple experts, Strictly confidential, Multiple bespoke

Table 5.4 Threat level as combination of the motivation and the capability according to TVRA

Motivation	Capability				
	Very little	**Little**	**Limited**	**Significant**	**Formidable**
Very low (indifferent)	Negligible	Negligible	Low	Low	Low
Low (curious)	Negligible	Negligible	Low	Low	Moderate
Medium (interested)	Negligible	Low	Moderate	Severe	Severe
High (committed)	Low	Low	Moderate	Severe	Critical
Very high (focused)	Low	Moderate	Severe	Critical	Critical

Target System In addition to describing the threat actor, motivation, and capability, we include the classification of the target system, i.e., whether the threat actor is primarily focused on compromising the vehicle or back-end servers, including internal IT and OT systems. The explicit inclusion of the target system can be helpful when working on different TARA projects, e.g., for different systems, by maintaining the same template or database for threat models. The capabilities of a threat actor may be different when analyzing embedded systems as opposed to IT/OT systems. In our view, separating target systems can be useful for the sake of better classification of different technology systems, such as distinguishing between a powerful back-end server cluster and a single vehicle telematics unit. However, with the increasing connectivity and complexity of vehicles, we can envision harmonization of vehicle and back-end systems in the execution of TARA. In fact, UN Regulation No. 155 already includes a list of threats that target back-end systems, even though the regulation focuses primarily on the vehicle [11]. We use the same attacker model described in this chapter for the case studies we present in this book for different target systems.

5.6.3 Relationship Between Attacker Model and Threats

The benefits of aligning the attack potential values of ISO/SAE 21434 with the threat levels as presented in Table 5.5 are significant. These values can be used to filter out specific threats that match the corresponding attack potential values or capabilities. This approach allows us to focus on evaluating the feasibility of attacks along specific attack paths related to specific assets. We can prioritize and focus on attack paths that require an attack potential capability and motivation level that exceed those defined in our attacker model for the specified threat level. For example, if our attacker model sets a maximum threat level of *Severe*, we can exclude from the TARA analysis those attack paths that require an attack potential equal to or greater than the combination of *Multiple Experts*, *Confidential*, *Bespoke*, and involve a highly motivated threat actor.

In addition, the inclusion of additional taxonomy criteria, such as target system, allows us to differentiate between threats that are solely focused on embedded devices (e.g., brute forcing a JTAG password or gaining elevation of privileges via UDS) and those that are focused on IT/OT systems. This distinction enhances the effectiveness of the TARA analysis by enabling a more targeted evaluation of relevant threats based on the specific characteristics of the system under analysis. Ultimately, a major benefit of having a good attacker model definition is that we can establish and follow a systematic, well-documented approach to performing TARA.

5.7 Practical Case Studies

We include three different case studies in this book that represent the most significant analyzes that are usually requested for a threat analysis and risk assessment. First, we analyze a single ECU. This represents the typical TARA project for embedded devices because ECU manufacturers need to perform a cybersecurity study, and of course, they are limited to protecting their product. As described previously and addressed by ISO/SAE 21434, the operational environment must be considered. In the second case study, we zoom out and analyze the connected vehicle in its digital ecosystem. We do not analyze the details of the internal architecture of the vehicle, although we consider the most relevant ECUs that can be used as an initial attack vector or target. In this case study, the scope of TARA is not limited to a specific ECU, as in the first case study, but to interactions with the outside world, such as

Table 5.5 Matching threat levels of TVRA to automotive threat actors, target systems and attack potentials

Threat level (TVRA)	**Threat agent (attacker)**	**Description**	**Targeting vehicles or back-end?**	**Motivation (TVRA)**	**Attack potential of threat actor**		
					Expertise	**Knowledge of the item**	**Equipment**
Negligible	Script kiddie	Someone who has a moderate know-how and skills and is able to find/buy code and applications online and can use them in a malicious way	Vehicle	Very low	Layman	Public	Standard
Negligible	Script kiddie	Someone who has a moderate know-how and skills and is able to find/buy code and applications online and can use them in a malicious way	Back-end	Very low	Layman	Public	Standard
Low	Tuner	Non-reliable workshop or professional tuner	Vehicle	Medium	Proficient	Public	Standard
Low	Data collector	Opportunistic hacker for data reselling	Vehicle	Medium	Expert	Public	Standard

(Continued)

Table 5.5 (Continued)

Threat level (TVRA)	Threat agent (attacker)	Description	Targeting vehicles or back-end?	Motivation (TVRA)	Attack potential of threat actor		
					Expertise	**Knowledge of the item**	**Equipment**
Low	Data collector	Opportunistic hacker for data reselling	Back-end	Medium	Expert	Public	Standard
Low	Thief	Car theft and break-in	Vehicle	Medium	Layman	Public	Standard
Low	Saboteur	Hacktivist, competitor	Vehicle	High	Expert	Public	Standard
Low	Saboteur	Hacktivist, competitor	Back-end	High	Expert	Public	Standard
Moderate	Motivated white-hat hacker	CCC member, IT-security professional or university researcher	Vehicle	High	Expert	Restricted	Specialized
Severe	Employee or ex-employee	Have insider know-how and trusted contacts in the company	Vehicle	High	Expert	Confidential	Specialized
Severe	Employee or ex-employee	Have insider know-how and trusted contacts in the company	Back-end	High	Expert	Confidential	Standard

(Continued)

Table 5.5 (Continued)

Threat level (TVRA)	Threat agent (attacker)	Description	Targeting vehicles or back-end?	Motivation (TVRA)	Attack potential of threat actor		
					Expertise	Knowledge of the item	Equipment
Severe	Hacker-for-hire actor	Illegal market hacking	Vehicle	Very high	Expert	Restricted	Specialized
Severe	Hacker-for-hire actor	Illegal market hacking	Back-end	Very high	Expert	Restricted	Specialized
Critical	Terrorist/ Cybercriminal	Highly skilled cyberterrorist teams or cybercriminal organization	Vehicle	High	Multiple experts	Confidential	Bespoke
Critical	Terrorist/ Cybercriminal	Highly skilled cyberterrorist teams or cybercriminal organization	Back-end	High	Multiple experts	Confidential	Bespoke
Critical	State-sponsored actor	APT, secret services	Vehicle	Very high	Multiple experts	Strictly confidential	Multiple bespoke
Critical	State-sponsored actor	APT, secret services	Back-end	Very high	Multiple experts	Strictly confidential	Multiple bespoke

mobile applications, cloud services, or location services. Finally, in the third case study, we analyze the risks at a supplier's manufacturing plant.

5.7.1 Case Study 1: Automotive Embedded Device

The first of our case studies is an automotive-embedded device, which is an electronic control unit (ECU) of a vehicle. We have chosen to analyze a modern head unit, also known as an infotainment unit. We are considering an advanced head unit with connectivity capabilities that replaces traditional telematics units. The head unit allows us to analyze an example of a classical ECU, but with advanced connectivity interfaces, such as USB, GNSS, and Wi-Fi, and also with powerful system-on-chips (SoC) for connectivity and video processing functions. The head unit we present is, of course, a simplified version of a real generic head unit for the didactic purposes of this book. A modern head unit can have multiple microcontrollers and microprocessors for different functionalities, including audio and video processing capabilities. Connections to the vehicle may include multiple CAN, FlexRay, and LIN buses. Human machine interfaces (HMI) such as touch screens, various buttons, and displays, are also part of a head unit. Additional comfort elements such as ambient light sensors, temperature sensors, and various cameras are also found in modern head units. The scope of our first case study is summarized in Table 5.6.

Table 5.6 Scope definition of our Case Study 1

General Description	
Item:	Head unit ECU
In scope:	Internal components of the ECU and interfaces to the outside
Operational environment:	The head unit interacts with the rest of the vehicle and with the outside world. It has a cellular interface for offering Internet access to the vehicle, Wi-Fi and Bluetooth interfaces for connecting the vehicle with a cell phone or a mobile device. (See complete description below)
Out of scope:	External components such as other ECUs or cell phones are not in scope. Only the data exchange from and to those are in scope
Impact categories:	Road User: Safety, Financial, Operational, Privacy OEM: Intellectual Property, Image, Financial, Operational
Attacker Model	
Threat level:	Attackers have a threat level *Severe*
Assumptions	
There is no previous TARA for this project	
No cybersecurity controls planned	

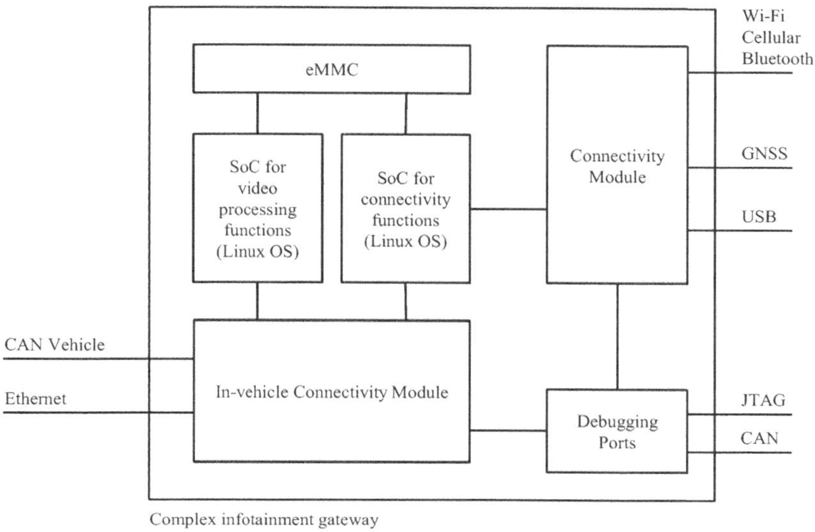

Fig. 5.3 Simplified E/E architecture diagram of our exemplary head unit

Item TARA focuses on the individual ECU. In scope are the interfaces to the outside of the ECU, i.e., all possible data traffic going in and out of the ECU. The item is defined as the head unit including external interfaces, as shown in Fig. 5.3.

In Scope The internal components of the ECU, such as connectivity modules, various SoCs, and memory slots, are in scope.

Operational Environment The head unit interacts with the rest of the vehicle and the outside world. It has a cellular interface to provide Internet access to the vehicle, Wi-Fi and Bluetooth interfaces to connect the vehicle to a cell phone or mobile device. Applications running on the ECU can access the Internet, for example, to request entertainment or vehicle-related services from the OEM. There is also a GNSS interface to provide location services to the vehicle. The driver or passenger uses navigation services, and the other ECUs can request location services as needed. A USB interface allows the ECU software to be updated and external hard drives to be connected for reading data for entertainment purposes. The head unit has a JTAG interface and a dedicated CAN bus for diagnostic purposes. The rest of the vehicle is connected to the head unit via an in-vehicle Ethernet network and a CAN network. The head unit can request operational data from other ECUs in the vehicle and provide diagnostic services including software updates. The head unit receives real-time video data from cameras outside the vehicle to record video data in the event of an accident. In addition, the user can remotely access the vehicle's cameras from a cell phone using a special application.

Out of Scope Components external to the ECU are out of scope. Only the data exchange to and from them is in scope. For example, we analyze the services provided to external cell phones and applications, but we do not analyze the internal workings of the cell phone or back-end servers in the scope of this TARA.

Impact Categories Apart from the four categories required by ISO/SAE 21434 concerning the road user, we consider the impact from the point of view of the OEM, which affects the categories *intellectual property*, *image* (in the sense of reputational damage), *financial*, and *operational*.

Attacker Model We limit the attacker model to the threat level *Severe*, as described in Table 5.5.

Assumptions We assume for this TARA that there has been no previous TARA and that the system has no planned cybersecurity countermeasures, at least at the time of our analysis.

5.7.2 Case Study 2: Vehicle-Backend Architecture

In our second case study, we analyze the entire connectivity ecosystem of a connected car. Rather than analyzing the inner workings of individual ECUs, as we did in Case Study 1, we analyze the vehicle as a whole. We model a vehicle with a simplified in-vehicle architecture. Furthermore, we do not analyze the in-vehicle architecture, but we are interested in attack paths that result in a compromise of the vehicle's ECUs. For this reason, we simplify the architecture to a head unit that mostly provides external connectivity, e.g., via the Internet, which is connected via a gateway to the rest of the classic ECUs and the OBD-II connector. The car is connected to cloud services to access OEM and third-party services. A cell phone allows access to the same cloud services as for the vehicle and in addition to connect directly to the vehicle. Location services are also included. The scope of this case study is summarized in Table 5.7.

Item The vehicle as a connected device and all possible devices connected to it, such as a mobile device or diagnostic tools, together with the back-end servers (i.e., cloud services) are in the scope of this TARA. Data and data handling are of particular interest to this TARA. The cell phone applications developed by the OEM or third-party companies that exchange data with the vehicle are in scope. The cell phones and servers themselves, i.e., their internal electronics and risks, are not and cannot be in scope. However, we can provide cybersecurity goals and requirements that such devices must meet to securely handle our data. The architecture of the element is shown in Fig. 5.4.

In Scope The scope includes location services, OEM and third-party back-end services, OEM and third-party applications, and diagnostic tools.

Table 5.7 Scope definition of our Case Study 2

General Description	
Item:	Connected car architecture, including cloud services and cell phone applications
In scope:	Location services, back-end services from the OEM and third-party companies, applications from the OEM and from third-party companies, as well as diagnostics tools, are in scope
Operational environment:	Diagnostics services including usage data; navigation; entertainment services such as music and social media; third-party services that can be installed into the vehicle such as a music or audiobook application. Payment services for OEM and third-party applications are present and managed by user profiles
Out of scope:	The internal workings of communicating systems are out of scope
Impact categories:	Road User: Safety, Financial, Operational, Privacy
	OEM: Intellectual Property, Image, Financial, Operational
	Third-party provider: Intellectual Property, Image, Financial, Operational
Attacker Model	
Threat level:	Attackers have a threat level *Severe*
Assumptions	
There is no previous TARA for this project	
No cybersecurity controls planned	

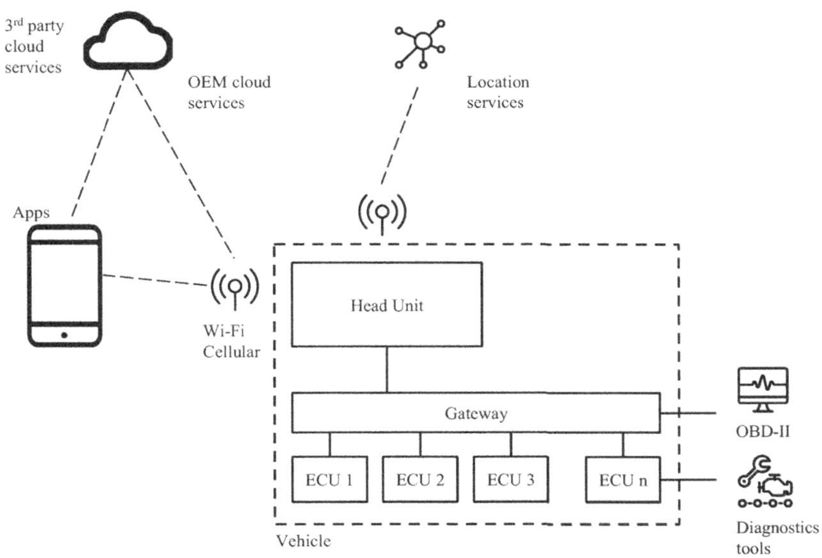

Fig. 5.4 High-level connectivity diagram of a vehicle-backend architecture

Operational Environment We consider the following remote services for the vehicle: diagnostic services that are provided by the vehicle and can be requested remotely, e.g., to obtain the current mileage, vehicle fault codes, and other usage data; navigation; entertainment services such as music and social media; and third party services that can be installed in the vehicle, like a music or audiobook application. The whole system is based on user profiles that are managed by the OEM and can be used for third-party services. Payment services are present and managed by the user profile so that an OEM or third-party application can use them to purchase services on-the-fly when requested by the user.

Out of Scope In general, the communicating elements such as the cell phone, the vehicle, the back-end servers are treated as black boxes for the sake of TARA and their internal workings are out of scope.

Impact Categories We analyze the four categories required by ISO/SAE 21434 for the road user together with new categories for the OEM and a third party. The categories for the last two cases are *intellectual property*, *image* (in terms of reputational damage), *financial*, and *operational*.

Attacker Model We limit the attacker model to the threat level *Severe*, as described in Table 5.5.

Assumptions Similar to our first case study, we assume there is no prior TARA and no planned countermeasures.

5.7.3 Case Study 3: Automotive Manufacturing Plant

In our third case study, we perform a TARA according to the procedures of ISO/SAE 21434 for a production environment. The production plan is a mixed IT/OT environment, and we apply concepts from the information security management standard ISO/IEC 27000-series and the OT cybersecurity management standard IEC 62443. The production plant can be, for example, for the production of the head unit ECU from our Case Study 1. The goal of the TARA is to identify the risks of introducing vulnerabilities during the production of the ECU and in the post-development (at least, threats that originate from or because of the backend/production systems). Figure 5.5 shows the high-level architecture of our production plan. It is a simplified form of a modern manufacturing plant, including a production line with internal and external connections. The production line has various machines. For example, a software flashing station that connects to a local server to obtain the software binaries to be flashed. In addition, the software flashing station connects to a public key infrastructure (PKI) system to verify digital signatures and install certificates and keys. Local monitoring and configuration allows operators in the field to monitor the

Table 5.8 Scope definition of our Case Study 3

General Description	
Item:	Manufacturing plant (IT/OT system) of a head unit ECU
In scope:	ECU being produced and the production plan, including machinery at the production line, and internal and external servers
Operational environment:	The ECU under production goes to the production line with different machines for different use cases, such as electrical or thermal testing. The first station is able to write the software contained in binary files obtained from a local server. The local server has a connection to the supplier's R&D department and to the OEM's back-end server to receive the information and software packages to be flashed into the ECU. A local PKI, or key management system, is used to validate and write certificates and keys into the ECU. The PKI, or key management system, connects to the OEM's key management system to obtain certificates and keys from the OEM, validate them, and synchronize relevant information. Both local and remote monitoring and configuration services are available
Out of scope:	Internal threats to machinery and servers, especially hardware threats, are not in scope
Impact categories:	Road User: Safety, Financial, Operational, Privacy Supplier: Intellectual Property, Image, Financial, Operational
Attacker Model	
Threat level:	Attackers have a threat level *Severe*
Assumptions	
There is no previous TARA for this project	
No cybersecurity controls planned	

status of the production line and make changes. It is possible to access the monitoring and configuration services from the Internet. The local server is also connected to cloud or backend services to download software packages from the R&D department located outside the production plant or directly from the OEM. The PKI or key management server is connected to an OEM key management server to download OEM certificates and keys and to synchronize the status of certificates and keys written during production. The scope is summarized in Table 5.8 and described as follows.

Item The item of the TARA is the production plant with its production lines, including internal and external servers and services. The objective of the TARA is to protect the ECU, which is the product being manufactured, from vulnerabilities introduced during its production. The production line is considered to be at an advanced or final stage of production, meaning that the ECU is physically built and ready to receive the final software.

In Scope The analysis is performed at an architectural level similar to Case Study 2, that is, we focus on the services running on the production facility, but not on the internal workings of each individual component. Of course, unprotected direct physical access to servers and interfaces is in scope, but not, for example, opening a

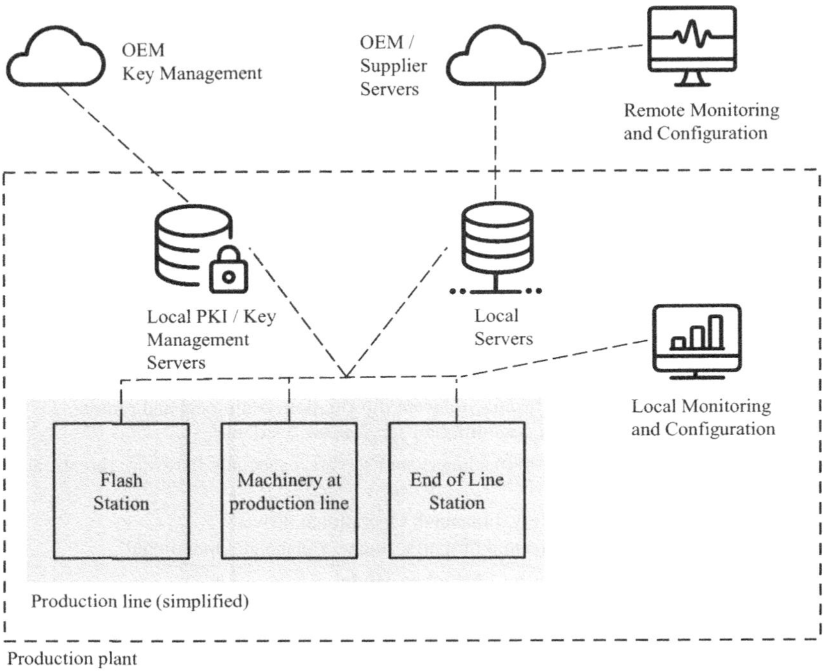

Fig. 5.5 Production plant including a production line and diverse back-end communication systems

server chassis and reading data directly from the server's RAM blocks. The ECU being built is in scope. Relevant threats are those directly related to the manufacturing plant. Other threats, such as compromising the ECU through connected CAN busses, are out of scope because they are covered by Case Study 1.

Operational Environment The ECU under production goes to the production line with different machines for different use cases, such as electrical or thermal testing. The first station is able to write the software contained in binary files obtained from a local server. The local server has a connection to the supplier's R&D department and to the OEM's back-end server to receive the information and software packages to be flashed into the ECU. A local PKI, or key management system, is used to validate and write certificates and keys into the ECU. The PKI or key management system connects to the OEM's key management system to obtain certificates and keys from the OEM, validate them and synchronize relevant information. Both local and remote monitoring and configuration services are available.

Out of Scope Threats against the internal hardware operation of the machines in the production plant are out of scope. The rationale behind this decision is that TARA should be able to provide cybersecurity goals to the production machines or servers that threaten the ECU. For example, if one of the goals is that the software should not be tampered with during the production line, then the team or supplier providing the machines for the production line must consider this goal and protect the machines against any tampering that could threaten the software binary on its way to the ECU.

Impact Categories Similar to our previous case studies, we analyze the four categories required by ISO/SAE 21434 for the road user, along with new categories for the supplier's perspective: *intellectual property*, *image* (in terms of reputational damage), *financial* and *operational*.

Attacker Model We limit the attacker model to the threat level *Severe*, as described in Table 5.5.

Assumptions We assume there is no prior TARA and no countermeasures planned.

References

1. ISO/SAE 21434:2021 Road vehicles—Cybersecurity engineering. Standard, International Organization for Standardization/Society of Automotive Engineers, Geneva, CH (2021). https://www.iso.org/standard/70918.html
2. Position—Automotive Cyber Security—Recommendation for Cyber Security Interface Agreement (CSIA), with reference to ISO/SAE 21434. Position paper, German Association of the Automotive Industry (VDA) (2021). https://www.vda.de/dam/jcr:8864645a-5dd8-4c16-b81c-708b0fc86b5c/VDA_Positionspapier%20CSS%202021_FINAL.pdf
3. Hug, N., Gibson, C., Kropotov, V., Vosseler, R.: Cybersecurity for Connected Cars (2021). https://documents.trendmicro.com/assets/white_papers/wp-cybersecurity-for-connected-cars-exploring-risks-in-5g-cloud-and-other-connected-technologies.pdf
4. MarketsandMarkets (2022). https://www.marketsandmarkets.com/Market-Reports/artificial-intelligence-manufacturing-market-72679105.html?gclid=CjwKCAjwu5yYBhAjEiwAKXk_eEGyyDDyDGxj4w6aVmo0tlRh6CgBicWFUL8ZhVU1RjvkRGkAKZpIfBoCKyUQAvD_BwE
5. Shostack, A.: Threat Modeling: Designing for Security, 1st edn. Wiley Publishing (2014)
6. ETSI: ETSI TS 102 165-1 V5.2.5—CYBER; Methods and protocols; Part 1: Method and pro forma for Threat, Vulnerability, Risk Analysis (TVRA). Standard (2022). https://www.etsi.org/deliver/etsi_ts/102100_102199/10216501/05.02.05_60/ts_10216501v050205p.pdf
7. Mead, N.R., Hough, E.D., Stehney, I., R., T.: Security quality requirements engineering (SQUARE) methodology. Tech. rep. (2005). https://insights.sei.cmu.edu/documents/751/2005_005_001_14594.pdf
8. Upstream: Upstream 2022 Global Automotive Cybersecurity Report (2022). https://upstream.auto/2022report/
9. European Union Agency for Cybersecurity (ENISA): ENISA Threat Landscape 2022 (2022). https://www.enisa.europa.eu/publications/enisa-threat-landscape-2022

10. Lautenbach, A., Mafijul Islam, C.: HEAling vulnerabilities to ENhance software security and safety (HEAVENS)—deliverable d2—security models (2016). https://autosec.se/wp-content/uploads/2018/03/HEAVENS_D2_v2.0.pdf
11. United Nations: UN Regulation No. 155—Cyber security and cyber security management system E/ECE/TRANS/505/Rev.3/Add.154 (2021). https://unece.org/transport/documents/2021/03/standards/un-regulation-no-155-cyber-security-and-cyber-security

Chapter 6
Identifying the Assets of the Item

Abstract In this chapter, we lay the groundwork for a TARA by identifying the assets of the system under analysis, or item. We provide practical insights into asset identification and discuss relevant cybersecurity properties in addition to the CIA triad. In addition, we provide a valuable list of common assets for both automotive embedded and IT/OT systems. Following asset identification, we move on to identifying damage scenarios. We analyze the damage scenarios affecting the road user and provide practical examples of how to evaluate the impact across the categories required by the standard. Following the ISO/SAE 21434 standard, we show how we can extend the view of the damage scenario to analyze the impact on an organization as a whole. For the business impact, we describe new impact categories, such as intellectual property and image/reputation. At the end of the chapter, we provide a complete asset identification for three practical case studies.

After defining the scope of the TARA, i.e., the configuration of the TARA and the information about the item, we proceed with the identification of the assets. The analysis of the assets can be divided into four high-level tasks. First, the *identification of the assets* of the item in the scope of the TARA that is used for the risk analysis. Second, the assignment of the *cybersecurity properties* to the assets, which practically means the analysis of which cybersecurity properties are relevant in case of an attack. Third, the identification of the *damage scenarios*, as established in the standard ISO/SAE 21434, and fourth, the evaluation of the *impact* (of a successful attack leading to damage scenarios) on different categories or high-level protection goals (e.g., safety, financial, operational, and privacy). These high-level steps are summarized in Fig. 6.1.

R. do Carmo and A. Schlensog, *Automotive Threat Analysis and Risk Assessment in Practice*, https://doi.org/10.1007/978-3-662-69614-9_6

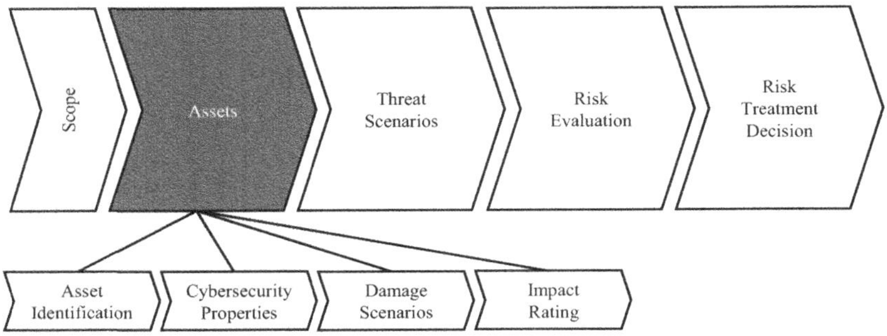

Fig. 6.1 High-level steps for the definition of the assets of the overall TARA process

Which step first: Asset Identification or Damage Scenarios?
ISO/SAE 21434 addresses the identification of assets (see *[RQ-15-02]* of [1]) immediately after the definition of damage scenarios (see *[RQ-15-01]* of [1]). However, this should not be seen as a rigid sequence, as both activities are interrelated. Furthermore, *Annex H* of the ISO/SAE 21434 standard even provides an example in which the order of the TARA steps or methods can be swapped. It is possible to start with the identification of assets and then move on to the identification of corresponding damage scenarios. Or it is possible to start with the selection of damage scenarios from catalogs and then move on to the identification of assets. In real projects, we have found that discussing the assets and their cybersecurity properties with the engineering team leads to a better definition of the damage scenarios. The need to define an asset only makes sense if it can be compromised, which ultimately leads to thinking about damage scenarios. Typically, this is a parallel activity, and the damage scenarios can be defined while the assets are being reasoned about.

Necessary input for this chapter:

- Technical documentation about the item and its operational environment (e.g., system architecture, technical diagram)

The outcome of this chapter is:

- Assets with cybersecurity properties
- Damage scenarios with impact rating

6.1 Definition of Asset

The definition of an asset can be as simple as an object that has value, as defined in [1]. For a TARA according to ISO/SAE 21434, the asset is something that has value and that has cybersecurity properties associated with it, such as confidentiality, integrity, and availability. The compromise of a cybersecurity property of an asset as a result of a cybersecurity attack can lead to one or more damage scenarios. In principle, any physical object or logical artifact such as data can be considered an asset. The ISO/SAE 21434 standard does not provide a catalog of assets, nor does it restrict what can be an asset. Depending on the scope of a TARA, the asset may be of a different nature. For example, when analyzing embedded devices, typical assets may include communication interfaces such as CAN or Ethernet, debugging interfaces, internal memory blocks, software components such as the bootloader or software applications, personal or sensitive data stored in the device, sensors, actuators, and others. When performing a TARA on a vehicle backend architecture or manufacturing plant, assets may include software applications, data on cell phones, databases on backend servers, payment information, and so on.

6.2 Asset Identification

According to ISO/SAE 21434, asset identification can be based on a variety of activities or sources, such as analyzing the item definition, performing an impact rating, deriving assets from threat scenarios, or using predefined catalogs (see *[RQ-15-02]* on [1]). In this section, we focus on analyzing the item definition in detail, which is the most comprehensive and natural way to identify assets. However, it is important to note that these activities and sources are not mutually exclusive, and new assets can be identified during subsequent steps of the TARA process, such as mapping threats or defining damage scenarios. Ultimately, the goal is to ensure a complete and thorough TARA that includes all relevant assets, threats, and risks.

In our experience, the most effective way to identify assets is through discussion with the engineering team and joint analysis of a system architecture diagram of the item. The analysis can begin with a functionality-based approach. The major functions of the item can be identified, along with their components, and the analysis can begin by essentially answering the questions *what could happen if the function or component is compromised?* and *how can the function/component be compromised?* The asset identification working group may typically include representatives from the following areas:

- *Cybersecurity:* The cybersecurity team, composed of one or more cybersecurity architects (or cybersecurity engineers, or similar cybersecurity-role names) coordinates the TARA activities and provides expertise on asset identification. The cybersecurity team is the main actor involved in the discussion about whether an asset is important or not from a cybersecurity perspective and for the scope of the

TARA (e.g., if it can be compromised as thought, if there are damage scenarios, if the asset and cybersecurity property are in the scope of the TARA)
- *Systems Engineering:* In cybersecurity analysis, it is essential to understand the big picture, i.e., how all the parts of the item under consideration interact. A systems engineer also provides useful information about the item's operating environment. This is especially important for reuse items, where a generic item or an item from another project is applied to a new project.
- *Hardware Architecture:* When performing TARA for embedded devices, hardware design engineers can provide useful insight into the overall hardware design and interfaces of the device.
- *Networking Engineering:* In today's connected vehicles, or when analyzing IT/OT systems in a manufacturing plant, the networking ecosystem plays a very important role. Network architects can help identify data flows and relevant network devices in the vehicle or plant.
- *Software Architecture:* The software team must be involved in a TARA to provide information about the software parts of the item. We emphasize that an in-depth analysis of the internal software development may not be necessary due to the simple fact that for cybersecurity, we analyze systems that are already working in the field. This means that the software is already compiled and running on the system. Typical relevant information for a TARA is the high-level architecture and distribution of software components, e.g., if there are configuration files stored separately from the binaries, if one or more bootloaders are running, if virtualization technologies are used, or how the software updates work.

A reasonable approach when involving different areas and teams is to divide the analysis into groups of experts because typically not all components are owned by the same people. For example, when analyzing an embedded device such as a head unit, the cybersecurity team leading the TARA can identify relevant functional components such as general interfaces, SoC for communication, SoC for video analysis, and so on, and invite the responsible people to different interview sessions. The asset identification steps with the support team are an iterative process until a consensus is reached between the cybersecurity team and the entire team. We depict these steps in Fig. 6.2[1].

[1] If the reader is currently inquiring about cybersecurity properties and damage scenarios, it is because these elements are integral parts of the overall asset identification process and cannot be separated from it. Later in this chapter, we will discuss cybersecurity properties and damage scenarios.

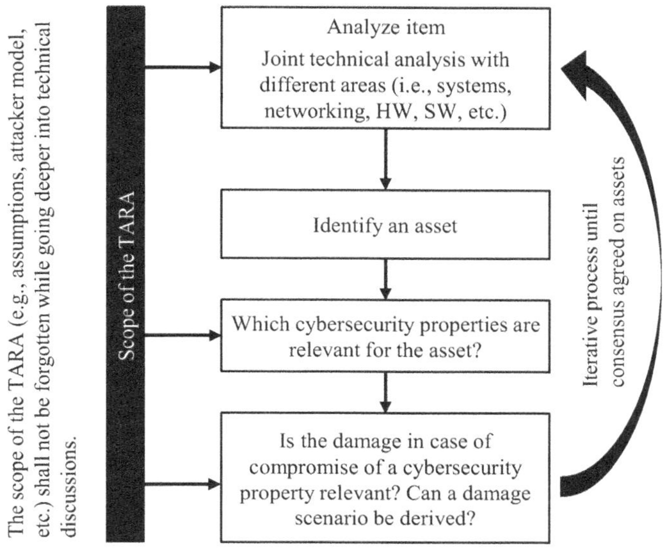

Fig. 6.2 Steps of the asset identification process

Tips

The cybersecurity team can start the asset identification discussion by following standard questions. The following questions are examples when performing TARA for an automotive embedded device:

- What interfaces does the device have, such as Ethernet or CAN?
- Does the device have wireless interfaces, such as Wi-Fi or Bluetooth?
- Are there any debugging interfaces, such as JTAG?
- If debugging interfaces are present, are they available in production? How are they protected?
- What diagnostic protocols are used? Are they standard diagnostic protocols, such as UDS, or proprietary protocols?
- Does the device have Internet connectivity? Is the device remotely accessible?
- Are any sensors or actuators connected directly to the device? If so, what protocol do they use to communicate (e.g., LIN, CAN, analog I/O)
- Are personal data processed and stored in the device?
- What data is processed by the device? Are logs generated? Is production/logistics data stored on the device?
- What software elements are present? Can the software be updated? How are software updates planned and via which interfaces are they carried out?

- Does the device store security-related data or artifacts, such as certificates and keys?
- Who are the communication partners of the device (e.g., other devices in the vehicle)?
- Does the device process safety-relevant data?

Asset Catalogs The joint review can certainly be supported by asset catalogs. The current assets for a TARA depend entirely on the needs of the particular project. However, the use of an asset catalog can be used as a source of inspiration at the very beginning of the asset identification process. The cybersecurity team can analyze the assets from multiple catalogs and compare them to the item. If the cybersecurity team is not sure if an asset applies to the item, the asset can be discussed with the entire team later in the joint discussions. Asset catalogs may be available in the organization, for example, as a collection of older projects. ENISA provides a comprehensive asset taxonomy for smart cars in [2]. In addition, we provide in this chapter a kind of catalog of the most relevant assets for different systems.

6.2.1 Typical Assets for Automotive Embedded Devices

We group the typical assets for automotive ECUs into the following categories: *communication interfaces/protocols*, *security-relevant data*, *personal data*, *other data at connected cars*, *software components*, and *hardware components*.

Grouping of Assets According to their Level of Abstraction We emphasize that it is good practice to group the assets of a TARA according to their functional level of abstraction. This is the case, for example, when analyzing communication protocols. It is possible to have only CAN as a single asset, despite its applications, if the scope of the TARA is to focus on the entire CAN bus. In another way, we can specify the services or messages transmitted over CAN, such as UDS or safety messages, as single assets. In this case, these individual services can be analyzed separately, and ultimately cybersecurity properties and protection goals can be defined for them separately. The following list describes typical assets found in automotive embedded systems for reference. However, not all of them need to be used together in all cases.

6.2.1.1 Communication Interfaces/Protocols

CAN The Controller Area Network (CAN), or simply CAN bus, is the most popular communication protocol in today's vehicles. It is a serial, two-wire communication bus protocol that began development in 1983 and is standardized by ISO 11898 [3, 4]. CAN is used to interconnect most of the ECUs of the vehicle, or at least those that belong to a certain group, such as powertrain or body controller, and data is broadcast between all participating ECUs. CAN Flexible Data Rate (CAN FD) is the modern successor to CAN for higher data rates. ISO-Transport Layer (ISO-TP) is an extension for sending data packets over CAN [5]. Typical applications of CAN are the exchange of vehicle data such as speed, temperature, collision detection. The exchange of safety-related messages between ECUs, sensors and actuators; and diagnostic services. If the ECU within the scope of TARA has a CAN interface, the incoming data must be carefully analyzed.

FlexRay It is another popular communication protocol used in modern vehicles for faster data rates than CAN, up to 10 Mbps. It is standardized by ISO 17458-1:2013 [6]. It is used for time-sensitive systems such as ADAS, steering and braking systems, or infotainment communications. FlexRay is more expensive than CAN and LIN, so it is implemented for high-end systems [7].

LIN The Local Interconnect Network (LIN) is the least expensive of the automotive protocols with a maximum speed of 20 Kbps [7]. It is a single-wire, serial protocol used for non-critical, low data rate applications such as small sensors, sunroof control, seat control, door locking or control, lighting, and others. It is standardized by ISO 17987-1 [8].

MOST The Media Oriented Systems Transport (MOST) protocol was developed for high-speed multimedia applications in automobiles. The slowest MOST variant, MOST25, uses plastic optical fiber as the physical layer and supports speeds up to 25 Mbps. The fastest variant, MOST150, supports up to 150 Mbps and is implemented on Ethernet [9]. MOST may not be of interest to TARA because it typically does not allow access to the vehicle's microphone, cellular system, or traffic information [7].

Ethernet Modern vehicles are moving away from protocols such as MOST and FlexRay for high-speed communication to Ethernet. Ethernet consists of different protocols for different purposes running at different OSI layers. It provides speeds up to 10 Gbps using non-proprietary protocols and any chosen topology [7]. Diagnostics over IP (DoIP) [10], Scalable Service-Oriented Middleware over IP (SOME/IP) [11], and Audio Video Bridging (AVB)/Time Sensitive Networking (TSN) for multimedia applications are typical automotive Ethernet protocols running at layers 5 to 7.

OBD/OBD-II The on-board diagnostics (OBD) connector, also known as the diagnostic link connector (DLC), is an external connector found in many vehicles that communicates with the vehicle's internal network [7]. OBD is nowadays also referred to as OBD-II because it is the current version that has been mandatory in many countries since some decades. In this book, we do not distinguish between OBD and OBD-II. It is an important asset from a cybersecurity perspective because it provides ports for several protocols such as CAN and KWP2000/K-Line, which are also used for standard diagnostics and software updates. The Keyword Protocol 2000 (KWP2000) standardized by ISO 14230 [12] provides diagnostic services. The European variant is the K-Line standardized by ISO 9141-2 [13].

JTAG This is a fairly critical asset because it is a protocol that allows low-level debugging capabilities as well as downloading and uploading firmware to a chip. JTAG connectors are usually located on the chips. JTAG can be disabled by eFuses. Tools like JTAGulator can even help find hidden or broken exposed pads for JTAG [7].

Wi-Fi It is a tremendous convenience to allow in-vehicle Wi-Fi connectivity to cell phones for accessing the Internet or multimedia services, or even to external devices such as the home access point for downloading over-the-air updates [14]. However, Wi-Fi connections could be exploited and open up a significant attack surface that could give hackers access to all vehicle systems, including safety-related bus communications such as CAN [15]. There are also privacy risks associated with Wi-Fi. For example, the vehicle can be tracked by sniffing the service set identifier (SSID), or personal data can be compromised if the vehicle's Wi-Fi router is manipulated [16].

Bluetooth Like Wi-Fi, it is a powerful interface for connecting the infotainment system to a mobile device. In the worst case, an attacker could gain access to personal data or even execute malicious code on the infotainment unit [7, 14].

USB Any USB port on any device in the vehicle is potentially exploitable and should be carefully analyzed. A USB connection to the infotainment unit can be exploited to gain root shell access to the system and install unauthorized software, as already happened to a major OEM [17].

eSIM If the vehicle or system has an embedded SIM (eSIM), the services or data transmitted must be analyzed as a cybersecurity asset. In modern vehicles, a telematics unit typically has one or more eSIMs to connect the vehicle to the Internet via cellular connections such as 4G, LTE, and 5G. Modern services include real-time traffic information, cloud-based back-end server applications such as remote vehicle control [14].

Digital/Analog I/O There may be different digital or analog I/O interfaces for different purposes, such as reading sensor data and controlling actuators. Such interfaces may fall within the scope of a TARA and therefore be considered assets. Threats to such interfaces may have minimal impact.

GPIO, I2C, SPI If on-chip threats are in the scope of the TARA, general-purpose input/output (GPIO) connectors may be considered for asset identification. Inter-Integrated Circuit (I2C) buses are likely to be found on chips for interconnecting components such as different processor units or memory blocks. Serial Peripheral Interfaces (SPI) can also be found on GPIO connectors or on the board itself.

Physical interface or "data" as asset?
An interesting discussion that arises from interfaces, especially for TARA for embedded devices, is whether to distinguish between the physical interfaces and the data flowing through them. The reason for this is usually that when analyzing E/E diagrams, the asset definition starts from a high-level perspective and the analysts tend to treat the interfaces as a whole functional block. In this case, for example, *CAN* is identified as an ambiguous asset, often meaning the physical connector or wires and the data flows. We believe a reasonable approach is to treat physical and logical assets as a single entity. In the end, by mapping the threats, it is clear which part of the asset is compromised. There may be projects where differentiation is required, for example, to make a strict distinction between physical and logical threats. In this case, it is possible to have two assets for CAN: *CAN Interface* for the physical interface, and *CAN Data* for the logical layer.

6.2.1.2 Security-Relevant Data

Public Keys and Certificates When identifying cybersecurity-relevant data, public keys and certificates are typical examples of data that must be defined as an asset because they have different cybersecurity properties than, for example, private keys, and they are an attractive target for threats. For this reason, tracking them as assets can help assign specific controls later in the TARA process. Public keys and certificates are typically not protected against confidentiality threats, since they are public by definition, but they are protected against integrity and authenticity threats.

Private Keys and Passwords Security secrets such as symmetric keys and passwords (i.e., JTAG passwords) must be handled carefully, since their disclosure often leads to harmful consequences.

Account Credentials Similar to private keys, if account credentials are present in the item, they can be analyzed separately as assets. Unlike keys and passwords, account credentials change over time, for example by changing drivers, and may be associated with other types of threats and controls.

Unique IDs OEMs and suppliers may use different types of identification numbers to identify parts of the vehicle and, in this case, security artifacts. Therefore, they must be considered in the threat analysis.

Secure Time on Vehicle Bus The transmission of time synchronization messages within the vehicle can also be an interesting asset and a target for various threats.

6.2.1.3 Personal Data

Driver/Passenger Personal Information Personally identifiable information (PII), such as name, address, birthday, biometric data, or any data that can be linked to a natural person, processed, and stored in the item, is a typical example of an important privacy-sensitive asset. Not only do automakers have to comply with strict privacy laws, such as the General Data Protection Regulation (GDPR) in Europe [18], but personal data is now one of the most valuable pieces of information an attacker or attacking group might be interested in. Leaking personal data can cause serious damage to the individual and the manufacturer [15].

Payment Information If the vehicle or item processes payment information, it can be the target of disclosure or manipulation, and therefore it is recommended to include it as a separate asset. Payment information can cause financial harm to the owner, for example, by revealing credit card numbers or making unauthorized transactions. In addition, it can be linked to PII and have significant privacy implications (e.g., by revealing name, address, or buying habits).

Geolocation Data The geolocation of the vehicle is an asset that should be included in the analysis of TARA, as it can reveal a person's elevator habits. If the journeys of the vehicle are tracked and stored, they can be misused to reveal sensitive information such as religion, sexual orientation, or other habits through the places visited [16].

Vehicle Identification Number The vehicle identification number (VIN) can link a vehicle to a natural person and is therefore a privacy-sensitive asset [16].

Video Data Modern connected vehicles may have different cameras for different reasons. For example, autonomous or advanced driving systems that use cameras for navigation, or dashboard cameras that record car accidents. There are even vehicles with cameras in the cabin to provide automated taxi services. While the exact scope of protection and requirements for video data may depend on the regulations of the country in which the vehicle is operating, video data can generally be considered PII and therefore a privacy-sensitive asset for a TARA.

6.2.1.4 Other Data at Connected Cars

Diagnostic Trouble Codes (DTCs) Trouble codes are five-character alphanumeric codes stored by the vehicle's electronic control units to provide diagnostic information. They can be in random-access memory (RAM) and stored in non-volatile memory to survive a power failure. Even if the DTCs only provide fault information, tampering can have negative consequences for the vehicle. For example, a tampered ECU can avoid writing critical DTCs, such as engine failure, and the driver can continue driving unaware of the problem. DTCs could also be spoofed, providing false information to the driver or vehicle owner [7].

Calibration Data Some ECUs can store variable data used for calibration purposes or for variant coding. This data must not be tampered with by unauthorized parties, so it must be protected.

Logs Any kind of logging information must be protected, at least against unauthorized manipulation. There may be cases where logs also store personal information. In this case, logs must be considered privacy-sensitive assets.

Mileage Counter Disconnecting, resetting, or altering the odometer of a motor vehicle, usually to show a lower mileage, is called odometer fraud and is a serious crime. However, in Germany alone, two million vehicles are tampered with every year [19].

User Profile Data User profile data can be present in some ECUs for various purposes, such as remembering seat positions, vehicle temperature, or other infotainment settings. In some cases, it may include personal data.

Life cycle State ECUs, and especially ECU microcontrollers, store data that represents the development state of the product. This information, typically referred to as the device life cycle, can be used to distinguish between the development, production, and in-field states of a product. Security features such as debugging interface protection or the use of generic passwords and keys can be linked to the device life cycle.

Data on RAM Data in the random access memory of embedded devices can be leaked by an attacker physically connecting to the ECU or by malware reading the RAM. Depending on the sensitivity of the data stored in RAM, such as personal data, it may be necessary to treat RAM data as an asset of the TARA.

Remote Monitoring Data The collection of data to monitor cybersecurity attacks, threats, and vulnerabilities is required by UN Regulation No. 155 (see, for example, paragraph 7.2.2.2 in [20]). Protecting this data is essential because the attacker could modify the data to hide an attack and prevent the OEM from detecting it. The attacker could also simulate a malfunction to fool the OEM into believing that a particular attack is taking place in order to divert attention for other purposes, such as a larger, fleet-wide attack.

Event Data Recorder (EDR) Data for crash investigation, e.g., to understand the circumstances of a crash or to reconstruct a crash. These data are stored on the event data recorder before, during, and immediately after the collision and may be subject to regulatory restrictions regarding privacy (i.e., unauthorized reading) and tampering protection. For example, EU Regulation 2019/2144 [21] prohibits the storage of data that could be used to identify the individual vehicle, its owner or holder, such as the full vehicle identification number (VIN).

Data Storage System for Automated Driving (DSSAD) With the advent of highly automated driving vehicles, it is important not only to store technical information about the status of the vehicle before, during, and after a crash or accident, but also to determine who was steering the vehicle at the time of the accident. The data storage system for automated driving is designed to store this information, including Global Positioning System (GPS) location and time, activation of automated driving functions, taking over of the driving task by the driver, and other information [22]. Due to its privacy and legal nature, if DSSAD is present at the time, it may be necessary to analyze it as an asset.

6.2.1.5 Software Components

Bootloader The bootloader of an ECU, or part of an ECU, is typically stored in a different memory area than the rest of the software, and it can be vulnerable to security attacks. There are many threats associated with the manipulation of automotive bootloaders. For example, some parts of applications, such as diagnostics, can be manipulated to disable authentication mechanisms for diagnostic purposes. The attacker can gain write access to sensitive information such as the odometer. Other threats may include disabling monitoring and safety functions while triggering false functions and putting the vehicle in a safety-critical state [23]. To evaluate specific threats and specific controls for bootloaders, it may be helpful to distinguish them as an asset from the rest of the software.

OEM Applications Parts of the software on embedded devices may be provided by the OEM with their binaries and designated memory locations. Depending on the scope of the TARA and the protection goals for the software, OEM applications may be a relevant asset. The cybersecurity properties are usually integrity and authenticity, so the ECU only needs to run software that really comes from the OEM and, of course, has not been tampered with. Confidentiality may be required to protect intellectual property, e.g., by preventing reverse engineering techniques.

Supplier Applications More than 80% of a vehicle comes from automotive suppliers [24]. If the ECU has individual pieces of software that originate only from the supplier, it may be considered an asset for TARA. Special consideration should be given if the supplier software runs on lower levels of the ECU to provide HW abstraction. Special debugging capabilities for aftermarket analysis at the supplier must also be carefully analyzed.

Third-party Applications Especially interesting for infotainment systems, third-party applications are an important asset because of their obvious powerful attack surface. A compromised application can essentially compromise the entire vehicle.

Software Updates The software updates for the vehicle's ECUs must be protected and considered an asset. If the item of the TARA is a typical embedded ECU, the asset could be, for example, the software binaries received via its interfaces for updating. Finally, the asset software update would have controls to prove its authenticity and integrity. At the vehicle level, TARA can derive cybersecurity goals for the vehicle back-end ecosystem that handles the software update process. Software updates and cybersecurity are interdependent. On the one hand, cybersecurity evolves over time as new vulnerabilities are found. These vulnerabilities are typically fixed and patched as software updates. On the other hand, the software update itself must be secure. For this reason, the UN Regulation No. 156 [25] came in parallel with the UN Regulation No. 155 on cybersecurity, and it requires OEMs to have a Software Update Management System (SUMS) established over the vehicle life cycle.

6.2.1.6 Hardware Components

ECU Housing The complete ECU at the hardware level can be useful for analyzing the physical types of attacks on the system. The asset is, of course, quite generic. The specific threats will depend on the analysis of the system under study. However, if the focus of the TARA is not on the physical level, but some basic threats need to be included, then a generic hardware asset may be useful.

Sensors A vehicle can typically have dozens of different sensors to monitor different areas, including coolant temperature, oil pressure, vehicle speed, oxygen in the exhaust, fuel level, parking obstructions, rain, tire pressure, and so on. The compromise of many of these sensors can have serious consequences for the vehicle and the driver. Even tire pressure monitoring system (TPMS) sensors can be compromised, as described in [7, 26]. TPMS signals can be eavesdropped to track vehicles, or spoofed TPMS packets can be sent to a vehicle, spoofing dangerous pressure and temperature information. In the worst case, a vehicle's battery can be drained by flooding the sensors with activation requests [7].

Actuators Similar to sensors, if there are actuators in the system in scope, they may be considered assets depending on the scope of the TARA and the impact that tampering with them could have.

Non-Volatile Memory Earlier in this chapter, we described the assets related to data stored or handled in a vehicle. The focus of threats and TARA can also be the hardware elements where the data is stored, typically called non-volatile memory (NVM) for persistent data. Various threats may be present, ranging from side-channel attacks or fault injection to simply connecting to the physical interfaces of the storage elements. Therefore, including non-volatile memory as an asset can be useful for highlighting these threats and possibly planning specific controls.

Power Supply The power supply connector or wires may be included due to simple attack vectors such as physically cutting the wires. However, for embedded devices at the required level, complex attacks can be related to the power supply of the embedded device or parts of its components, such as microcontrollers. One powerful attack is the fault injection attack, also known as the voltage/power glitching attack [27, 28].

6.2.2 Typical Assets for IT/OT Systems

We present a list of more generic assets that are typically found in both vehicle backend architectures and manufacturing plants.

Debug/Diagnostic Interfaces Similar to embedded devices, IT devices and manufacturing machines can have different types of interfaces used to diagnose the device. The interface can be a standard interface, such as USB, and used together with other services, or it can be a special interface, perhaps following a proprietary hardware interface and/or communication protocol.

USB USB device interfaces can serve several purposes. Typically, they are used for connecting peripherals, such as a keyboard or mouse, but they can also be used for maintenance, such as booting from an external USB drive for firmware updates. It is recommended to carefully analyze the USB interfaces in TARA.

Wi-Fi There can be many reasons for having Wi-Fi on IT/OT systems. Due to its popularity and ubiquity, Wi-Fi interfaces are an obvious asset to analyze in detail.

Fieldbus Industrial computers and production machines, such as sensors, actuators, and PLCs are often connected using fieldbus protocols. Fieldbus consists of a family of protocols, classified as profiles, standardized by the IEC 61158 family of international standards [29]. Popular fieldbus protocols are, for example, MODBUS [30], ControlNet [31], and PROFIBUS [32]. According to [33], PROFIBUS is the world's most successful fieldbus with 65.9 million devices installed lately. Nevertheless, PROFIBUS and other fieldbus protocols are not safe from cybersecurity threats, as evaluated in [34].

Ethernet Ethernet interfaces, like Wi-Fi, are powerful and extremely popular, and they are an obvious asset to analyze in any TARA. In industrial systems, Ethernet can be used for many purposes, such as transferring production and logistics data, operating ICS devices such as PLCs, or transferring SW packages and security-related data to the product on the production line. Industrial systems typically have a long life cycle, and machines run for decades with minimal downtime. It is common today to find outdated or insecure protocols, such as SMBv1, running on industrial systems.

User Interface Any type of HMI can be used as an attack vector. There are many types of popular interfaces, such as touch panels, and their specific use and significance from a cybersecurity perspective will depend on the particular project and device being analyzed.

Remote Access IT/OT systems may have external connectivity over the Internet for various reasons. Manufacturers and administrators of industrial equipment may have remote diagnostic access to quickly resolve problems or provide software updates. External access can also be used for operational purposes, such as transferring software packages between sites, logistics data used in production, or even security-related data with the manufacturer and the customer. In addition to administrative and operational connectivity for the manufacturing plant, the company's office network can be connected to the production network, exposing the plants to classic IT threats such as malware via infected emails or files, phishing, social engineering, and others. In fact, according to [35], the infection of ICS systems with malware via the Internet and intranet has been a trending vulnerability recently.

Operating System Access to administrative parts of the operating system running on OT devices can open the door to a variety of different threats. Classic vulnerabilities of operating systems for ICS devices are poor administrative processes, i.e., ordinary users having full access to applications for their regular work, poorly chosen passwords, insecure update management or unprotected services [36]. For the purposes of TARA, it is recommended that the operating system be explicitly mentioned, as it may be subject to different threats and controls than the rest of the application software.

Applications The applications running on IT/OT systems can be used for various purposes and can be associated with significant cybersecurity risks. Applications can include control software for production machinery, management of production and logistics data, databases, and many others.

Software Updates The update software package and the update process itself can be a source of powerful attacks. If the integrity and authenticity of a software update is not guaranteed, a malicious software update can have direct access to software that controls data and physical systems. Even ransomware can be distributed throughout the manufacturing plant if it can infect a software update.

Logistics Data For example, in manufacturing plants, logistics data used to identify hardware and software parts, or to provide monitoring and traceability information, is transferred between different production lines and IT systems. Malicious modifications to logistics data could result in incorrectly produced products, such as debugging software instead of production software, or even halt production.

Bill of Materials (BOM) The bill of materials can be considered part of the asset's logistics data. We refer to it as a separate asset to emphasize its importance and criticality. Careless manipulation of the BOM in production, for example, can lead to the wrong software being flashed on the assembly line, or even to the assembly lines building the product (e.g., an ECU) with the wrong or missing parts.

Software Bill of Materials (SBOM) The software bill of materials is a formal record that contains the details of the software components used in a software build for a product. These components can be, for example, open-source libraries, proprietary libraries, modules, configuration parameters, interdependencies, and others. The SBOM plays an important role in cybersecurity, such as identifying and avoiding known vulnerabilities. With an SBOM, it is easier for OEMs, suppliers, and automotive software vendors to keep track of vulnerabilities discovered in software already in vehicles in the field and provide a patch. The integrity and authenticity of an SBOM must be ensured throughout the supply chain.

Security-relevant Data If the product being manufactured requires cybersecurity artifacts such as certificate signing requests (CSR), symmetric keys, or passwords, this data will be present in the industrial network of the manufacturing plant. It is likely that internal replicas of key management system servers are running inside the plant and exchanging information with external parties, such as the manufacturer's network or the customer's network. Security-sensitive data must be handled carefully to prevent manipulation and unauthorized access.

Sensors/Actuators While data flows to sensors and actuators within a production line may be covered by the assets associated with networks such as fieldbus or Ethernet, compromising the physical integrity of sensors and actuators may be of particular interest to TARA and within its scope as individual assets.

Servers A generic asset covering servers in the production plant, e.g., for transferring operational data or security-related data, may be useful for mapping threats to the HW and SW of classical IT systems.

Networking Devices Networking devices such as switches and routers are naturally exposed to cybersecurity threats, especially from internal attackers. Managed systems can be compromised from the outside if remote access is not properly configured and networks are not properly segregated.

End-Product/PCB The ultimate goal of a TARA from an ISO/SAE 21434 perspective is to protect the production of vehicles and their components. Within a manufacturing plant, the end product may be threatened by unauthorized access or direct physical manipulation. A direct connection to a printed circuit board (PCB) can allow a malicious operator on the production line to compromise its hardware or software.

Power Supply The power supply is a crucial asset in the overall operation of a manufacturing plant. While compromising the power supply may be most threatened by physical sabotage, it can be analyzed for a TARA if there are intelligent power management systems that may be exposed to cybersecurity threats.

6.2.3 Safety-Relevant Assets

Functional safety is a core discipline in the automotive industry. While there are harmonization efforts to integrate cybersecurity and functional safety, they are typically kept separate in organizations today. The ISO/SAE 21434 standard shares some relationships, especially in terms of definitions, with the ISO 26262 [37] standard for functional safety. For example, the term *cybersecurity goal* is related to the term *safety goal*, and the terms *item* and *item definition* even share the same name. Even the process used in the ISO 26262 standard for identifying and categorizing hazardous events, called Hazard Analysis and Risk Assessment (HARA), shares similarities with TARA.

When looking at the asset identification process at TARA, it is reasonable to consult safety experts working on the same project. For example, if a HARA already exists, the cybersecurity team can analyze whether the elements of the ISO 26262 item definition can be used as assets. In addition, the hazards identified in a HARA can be used as a source of inspiration for identifying assets. For example, if there is a hazard of *accelerating instead of braking due to errors in bus signals*, the cybersecurity team can include *bus signals* or *safety-related bus signals* as part of the assets. Ultimately, safety-related bus signals must function properly, regardless of an electromechanical failure or cybersecurity attack. The ISO/SAE 21434 standard even requires organizations to establish and maintain communication channels between relevant disciplines, including functional safety (cf. *[RQ-05-05]* at [1]). Safety parameters are also used later in the TARA when assessing the impact of damage scenarios (see Sect. 6.4.2.1).

Another relevant discipline is Safety of the Intended Functionality (SOTIF) as defined in the international standard ISO 21448 [38]. SOTIF differs from functional safety in that it focuses on hazards resulting from functional insufficiencies of the intended functionality or its implementation (rather than hazards resulting from malfunctioning of E/E systems). SOTIF has become particularly relevant with the advent of autonomous vehicles. A typical example of a situation analyzed by SOTIF is the misinterpretation of traffic signals by an autonomous vehicle, such as the misinterpretation of a traffic light. For example, the algorithm used to interpret the red color of a traffic light could be incorrectly programmed, or the sensor could not provide a clear image due to incorrect lighting conditions. The cybersecurity team performing the TARA can also conduct a joint review of the relevant hazards identified by SOTIF to derive assets for a cybersecurity analysis.

6.3 Cybersecurity Properties

According to ISO/SAE 21434, cybersecurity properties are the attributes of an asset that can be threatened (from a cybersecurity perspective) and, if compromised, can lead to a damage scenario. The concept of modeling security attributes of an asset to formalize cybersecurity risk management is not new. In fact, the attributes applied by ISO/SAE 21434 are the well-known CIA triad: confidentiality, integrity, and availability. The same concept of cybersecurity properties is found in the literature for different domains with different names, such as security attributes in HEAVENS [39] or cybersecurity goals. The IEC 62443 standard addresses confidentiality, integrity, and availability under the name of security objectives [40].

Assigning cybersecurity properties to the assets identified in this chapter is an essential step in TARA. Not all cybersecurity properties are relevant to all assets. The threats identified in the next step of the TARA are assigned to the assets. The cybersecurity properties compromised by the threats must match the cybersecurity properties that are important to the assets. As a trivial example, the two relevant cybersecurity properties for the odometer counter stored by an ECU may be integrity and availability. It is imperative that the odometer counter is not intentionally tampered with, for example, to prevent odometer fraud. And for operational use, it must be available when needed. In this case, only threats that compromise the integrity and availability of the odometer counter are considered for TARA. If an asset does not have a particular cybersecurity property, then there is essentially no threat to that asset. This may sound obvious, but it is important to keep these concepts in mind when performing TARA, especially when working with pre-populated catalogs such as threat catalogs.

6.3.1 CIA Triad: Confidentiality, Integrity, Availability

The properties of the CIA triad, confidentiality, integrity and availability, are used along the standard ISO/SAE 21434. These properties are well established in the InfoSec community, and therefore must be considered as the standard set of cybersecurity properties for a TARA project.

6.3.1.1 Confidentiality

Confidentiality is the protection of information from unauthorized access. In ISO/IEC 27000, confidentiality is defined as the property that information is not made available or disclosed to unauthorized individuals, entities, or processes [41]. Typical assets that must be kept confidential include anything related to the operation of cybersecurity protections, such as private keys and passwords, and assets related to sensitive and personal data, such as payment information or driver's names and home addresses. We recommend carefully analyzing whether confidentiality is a relevant cybersecurity property for an asset, or whether it can be omitted. A common control to protect the confidentiality of data at rest and in transit is encryption. Encrypting data can be burdensome for certain smaller ECUs with limited resources, and it can be prohibitive (i.e., without improving the HW) for systems that rely on real-time responses, such as for powertrain applications. Even with faster symmetric algorithms, there is still the technical challenge of distributing the required keys and keeping them protected. For this reason, encryption must be considered only when it is really necessary, and not in an inflationary way for assets that may only need integrity protection, for example.

6.3.1.2 Integrity

Integrity is the property of accuracy and completeness of information [41], which means that it should be possible to verify and know that information has not been modified in transit by unauthorized entities [42]. Data modification does not always have to be intentional. It is possible for data to be stored in a corrupted area of memory, or for data to be modified due to interference problems in certain transmission protocols.

6.3.1.3 Availability

Availability is the property of the asset to be available, accessible, and usable when needed. Because of its operational nature, availability is a property that is widely applied to many assets. For example, in a high-level system view, the control elements of a vehicle, such as the steering wheel and brakes, must be available at all times to ensure the proper and safe operation of the vehicle. Applied to embedded systems, for example, the software that controls the safety-related messages of an electronic braking system unit must have high availability. These units make decisions in milliseconds that affect the vehicle's dynamics. But availability is not just important for safety-critical components. Any type of information that is used at any point must be available when the user or interested entity needs it. Examples include user profile information for the infotainment system, Internet access for real-time traffic information, or error codes for diagnostic evaluation in the workshop. In manufacturing plants, for example, the availability of industrial control systems may have a higher importance and priority than other properties, as detailed in [40].

6.3.2 *Other Cybersecurity Properties*

In addition to the CIA triad, there are many other cybersecurity properties in the literature. For example, Parker [43] proposed three new properties to extend the CIA triad: possession or control, authenticity, and utility. HEAVENS proposes eight cybersecurity properties: confidentiality, integrity, availability, authenticity, authorization, non-repudiation, privacy, and freshness [39]. For the purposes of the TARA according to ISO/SAE 21434, the analysis of the CIA triad may be sufficient, but there is no restriction to include more cybersecurity properties in the TARA. Based on our professional experience in the field and for the sake of illustrating examples on other properties than CIA, we include in our methodology the additional cybersecurity properties *authenticity*, *non-repudiation* and *accountability*. Authenticity, however, is much more widely applied than non-repudiation and accountability.

6.3.2.1 Authenticity

Authenticity can be defined as the property of information or assets being genuine, verifiable, and trustworthy [44]. In other words, authenticity is the ability to be sure that data came from the expected sender or owner. The term should not be confused with *authentication*, which is the act of validating the identity of a user or entity. Authenticity is an important property in today's systems. For example, in embedded devices for vehicles, the ECU can verify that the code installed in the ECU was generated by the OEM or another valid provider. This can be achieved using public certificates and digital signatures. The ECU can also verify the authenticity of software updates received by the ECU before installing them.

6.3.2.2 Non-Repudiation

Non-repudiation is a classic property in the field of information security and cryptography, and consists of making it impossible for an individual or entity to falsely deny having performed an action [45]. In simpler terms, it prevents a sender from later denying that they sent a message or carried out a specific activity [42]. Non-repudiation is an important cybersecurity property in the context of modern connected vehicles. Consider the example of V2X communication, where vehicles exchange critical messages such as speed information, traffic conditions, or speed limit data. In such scenarios, it is critical to have a mechanism in place that uniquely identifies the sender of the message and ensures non-repudiation. By implementing non-repudiation measures, it becomes possible to hold the sender accountable for their actions and establish a level of trust in the communication system. When a sender is uniquely identified and their actions are non-repudiable, they can be held legally responsible for the information they transmit. This is particularly important in situations where critical decisions, such as autonomous vehicle operations or safety-critical notifications, depend on accurate and trusted data exchange [23].

6.3.2.3 Accountability

Accountability is the property that ensures that information or actions of an entity, such as a person or a system, can be uniquely traced to the entity that can be held responsible for that information or action [40, 46]. In the context of cybersecurity, accountability is often linked to enterprise-level policies as defined by standards such as IEC 62443 [40]. These policies outline areas of responsibility within an organization, such as IT and plant operations, and assign specific roles and obligations to ensure accountability. By clearly defining these responsibilities, organizations can establish a framework for accountability and enforce proper actions and behaviors. For instance, in the case of autonomous or highly automated vehicles, accountability becomes a significant legal challenge. Accidents involving such vehicles raise questions about who should be held liable and responsible. Establishing accountability in this context requires defining the roles and responsibilities of various stakeholders, including vehicle manufacturers, technology providers, and regulatory bodies.

6.4 Damage Scenarios

In the context of ISO/SAE 21434, a *damage scenario* represents an adverse consequence for a road user as a result of a cybersecurity attack. A damage scenario occurs when a cybersecurity property of an asset is successfully compromised by an attack. This means that the threat scenarios with their individual attack paths for the different cybersecurity properties of the assets converge into damage scenarios. Many threat scenario-asset combinations can result in the same damage scenarios. For example, manipulating different types of messages or signals over different communication channels, such as CAN and Ethernet, can lead to the same damage scenario defined as *the vehicle accelerates without driver intervention*. We recommend combining the damage scenarios and keeping the number of damage scenarios small. This is particularly practical in order to have a manageable list of damage scenarios for the road user, while having many possible threat-asset combinations associated with them. In this step, the impact of a successful attack is evaluated for each damage scenario. The relationship between attack paths, threat scenarios, assets, and damage scenarios is shown in Fig. 6.3.

Missing the Whole Picture Damage scenarios ultimately represent vehicle-level or system-level adverse consequences for road users. This means that damage scenarios

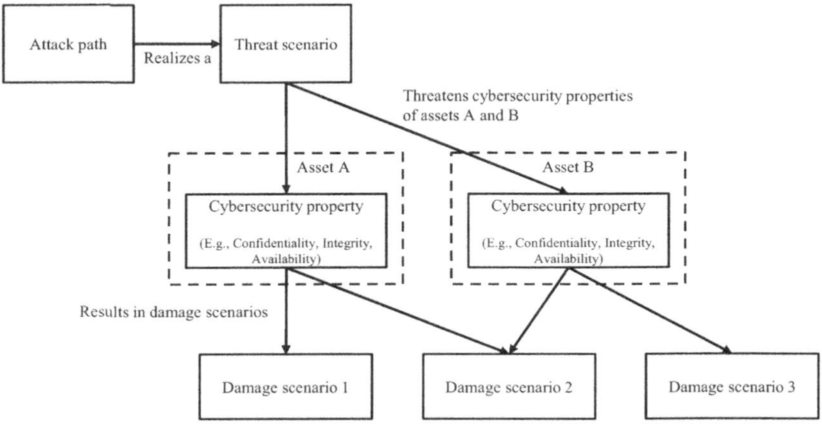

Fig. 6.3 Relationship between attack path, threat scenario, assets, and damage scenarios. A same threat scenario compromises two assets, Asset A and Asset B, which results in different damage scenarios

involve the whole picture of a vehicle. When performing component-level TARAs, it is sometimes difficult to establish a direct link from the component to the vehicle and thus to derive a damage scenario. In cases where there is a shared responsibility between the supplier and the OEM, such as where the supplier only provides a *generic TARA* for a generic out-of-context component, the damage scenarios may not be known to the supplier (cf. Sect. 5.3.1.1). Items, by definition, have vehicle functions associated with them, and the link to damage scenarios is clear (e.g., the ECU sending messages directly into a safety-related vehicle CAN). However, we also observe that suppliers responsible for conducting TARA cannot always link cybersecurity attacks to an item with high-level damage scenarios. Additionally, the whole picture can also involve mechanisms related to the operational environment of the item. For example, it is unlikely that safety-related messages are accepted by an ECU controlling brakes or steering, making the vehicle perform an action without regard to other safety conditions. In these cases, when performing TARA for items in a generic manner, or for components without a direct connection to vehicle functions, TARA can still be performed by evaluating the vehicle-level damage scenarios and assuming worst-case conditions. For example, tampering a component may lead to malicious operation of the item (assuming no other countermeasures are in place), which may ultimately affect a vehicle-level function.

6.4.1 Identification of Damage Scenarios

The identification of damage scenarios can be accomplished using the same brainstorming methodology for identifying assets as shown in Fig. 6.2. The compromise of any cybersecurity property of the assets defined for the TARA results in one or more damage scenarios. These scenarios can be identified through a brainstorming session with the expert team, where each cybersecurity property of each asset is evaluated for potential threats.

In addition, the cybersecurity experts supporting the TARA may already have experience with the asset, and a list of damage scenarios may already be available, for example from previous TARAs. Using a catalog of damage scenarios can provide a useful baseline for comparison, especially if similar TARAs already exist for the same item. However, it is important to exercise caution when using these catalogs. While they can be a valuable resource, relying solely on them could lead to overlooking damage scenarios that are specific to the item and its identified assets. Therefore, it is crucial to use these catalogs as a starting point, but also to engage in thorough brainstorming and analysis to identify any unique or item-specific damage scenarios.

UN Regulation No. 155 [20] provides in its *Annex 5* an exemplary list of possible attack impacts. These attack impacts can serve as a source of inspiration when discussing damage scenarios, providing a broad range of potential impacts to consider. However, as with the use of catalogs, these should be used as a guide rather than a definitive list, and the unique characteristics and context of the item should always be taken into account.

Attack Impacts from UN Regulation No. 155

The following attack impacts are listed in Annex 5 of UN Regulation No. 155 [20]:

- Safe operation of vehicle affected
- Vehicle functions stop working
- Software modified, performance altered
- Software altered but no operational effects
- Data integrity breach
- Data confidentiality breach
- Loss of data availability
- Other, including criminality

It should be noted that they are not directly called "damage scenarios", nor are they related to the road user as in ISO/SAE 21434, but they can be considered for the formulation of damage scenarios.

6.4.2 *Impact Rating*

ISO/SAE 21434 specifies that the impact rating of the damage scenarios shall be assessed against at least the categories *safety*, *financial*, *operational*, and *privacy*

Table 6.1 Safety impact rating criteria for road users (as in Table F.1 of [1])

Impact rating	Criteria for safety impact rating
Severe	S3: Life-threatening injuries (survival uncertain), fatal injuries
Major	S2: Severe and life-threatening injuries (survival probable)
Moderate	S1: Light and moderate injuries
Negligible	S0: No injuries

(see *[RQ-15-04]* of [1]). Additional categories are allowed, and we analyze further categories in this chapter. The rating has four levels for each category: *negligible*, *moderate*, *major*, and *severe*. The rating criteria can be defined by the organization. In this case, it is important to properly document the criteria and the rationale behind them. In this section, we follow the example of impact criteria provided in *Annex F* of ISO/SAE 21434 for road users.

6.4.2.1 Safety

The impact rating value and criteria for the road user view shown in Table 6.1 are taken from the well-established international standard ISO 26262 [37] for functional safety of road vehicles. The security team performing the impact rating is not expected to be familiar with the functional safety standards and procedures used in the company. For example, a joint review with functional safety experts may be useful to gain a more profound understanding of which impact rating better matches the current analysis for the damage scenario. Other functional safety details, such as the controllability and exposure parameters from ISO 26262, can also be included in the criteria definition.

6.4.2.2 Financial

The examples of financial impact criteria for road users from ISO/SAE 21434 are shown in Table 6.2. The definitions are, of course, quite general. How to differentiate or apply practical examples between inconvenient, substantial and catastrophic consequences for the moderate, major, and multiple categories may depend on the particular project and company-specific rating descriptions. For example, a moderate loss could be a vehicle that is not fully operational, causing the driver or passenger to lose time and money. The driver or passenger may decide not to use the vehicle and take a taxi to get to their appointment. A moderate impact could be the loss of sensitive information from the road user, such as bank account details and payment information. Severe impacts could include pedestrians being injured by a vehicle, hospitalization costs, and loss of employment.

Table 6.2 Financial impact rating criteria for road users (as in Table F.2 of [1])

Impact rating	Criteria for financial impact rating
Severe	The financial damage leads to catastrophic consequences, which the affected road user might not overcome
Major	The financial damage leads to substantial consequences which the affected road user will be able to overcome
Moderate	The financial damage leads to inconvenient consequences which the affected road user will be able to overcome with limited resources
Negligible	The financial damage leads to no effect, negligible consequences, or is irrelevant to the road user

Table 6.3 Operational impact rating criteria for road users (as in Table F.3 of [1])

Impact rating	Criteria for operational impact rating
Severe	The operational damage leads to the loss or impairment of a core vehicle function
Major	The operational damage leads to the loss or impairment of an important vehicle function
Moderate	The operational damage leads to partial degradation of a vehicle function
Negligible	The operational damage leads to no impairment or non-perceivable impairment of a vehicle function

6.4.2.3 Operational

Table 6.3 shows the criteria examples for operational impact to the road user. ISO/SAE 21434 provides some examples of what moderate, major, and severe can be. For example, a level of moderate can be a negative impact on user satisfaction. A level of major can be a significant annoyance to the driver. And the highest level, severe, can be a complete loss of functionality of a component or the entire vehicle.

6.4.2.4 Privacy

The criteria for rating the impact on the privacy of the road user are shown in Table 6.4. Some definitions in the criteria levels are taken from the ISO/IEC 29100 standard [47]. In particular, the following definitions are relevant to the understanding of the table.

Personally identifiable information (PII): Any information that (a) can be used to identify the PII Principal to whom such information relates, or (b) is or could be directly or indirectly linked to a PII principal [47].

Table 6.4 Privacy impact rating criteria for road users (as in Table F.4 of [1])

Impact rating	Criteria for privacy impact rating
Severe	The privacy damage leads to significant or even irreversible impact to the road user The information regarding the road user is highly sensitive and easy to link to a PII principal
Major	The privacy damage leads to serious impact to the road user The information regarding the road user is: a) highly sensitive and difficult to link to a PII principal; or b) sensitive and easy to link to a PII principal
Moderate	The privacy damage leads to inconvenient consequences for the road user The information regarding the road user is: a) sensitive but difficult to link to a PII principal; or b) not sensitive, but easy to link to a PII principal
Negligible	The privacy damage leads to no effect or, negligible consequences or is irrelevant to the road user The information regarding the road user is not sensitive and difficult to link to a PII principal

PII principal: The natural person to whom the personally identifiable information (PII) relates [47].

Examples of PII include: name, gender, date of birth, medical history, biometric identifiers, sexual orientation, credit card statements, home address, GPS location, GPS trajectory, work address, personal or behavioral profile, personal interests derived from tracking use of Internet sites, and criminal investigation reports. For an exhaustive list, see [47].

6.4.3 Additional Stakeholders

The analysis of the impact on other stakeholders than the road user is not provided for in ISO/SAE 21434. However, in addition to the road user, it may be useful to analyze the impact on other subjects to get a more holistic approach to TARA and risk management. In practical TARA projects, it is usually interesting to analyze the consequences of damage scenarios on the OEM or supplier. The damage scenarios affecting the OEM or the supplier may be the same as those affecting the road user. In addition, there may be damage scenarios that only affect the OEM or the supplier. In this case, it is recommended that these damage scenarios be described or given a special name to separate them from the rest of the TARA for compliance purposes.

Not all categories that apply to road users are meaningful when analyzing the impact on OEMs or suppliers. For example, safety impacts may be irrelevant to the analysis of OEMs and suppliers as companies as a whole. For example, if the analysis is focused on production facilities and the safety of personnel working in the production environment needs to be considered, then the safety category will become relevant. In the following examples of impact categories, we analyze OEMs and suppliers as a whole company or entity, so we do not use the categories *safety* and *privacy*. Instead, we suggest the categories of *intellectual property* and *image or reputation*. We keep the *financial* and *operational* categories, but adapt the criteria to better fit our use case. The criteria are based on our experience. In addition, we have taken inspiration from IEC 62443-2-1 [48].

6.4.3.1 Intellectual Property

We include intellectual property in our impact analysis because it is one of the most valuable assets of modern companies. In fact, intellectual property can make up to 80% of the value of a simple company [49]. It is possible and valid to define intellectual property as an asset according to the TARA asset definition and then map threats to it. The advantage of using intellectual property as a category for impact analysis is that it may be easier to identify the actual threats to intellectual property, since the category must be evaluated orthogonally for each combination of threat and damage scenario. In addition, the importance of intellectual property is highlighted when it is an impact category. In Table 6.5, we describe the different criteria for assessing the impact of intellectual property that we propose in this book.

Table 6.5 Intellectual property impact rating criteria for OEM or supplier

Impact rating	Criteria for intellectual property impact rating
Severe	Catastrophic impact for the company, leading to costly investments in research and development and loss of sales
Major	The disclosure of these data might lead to significant damage for the company Competitors might take advantage of this information. For example, engineering details on the system design, system architecture, or even internal organizational information
Moderate	Secret data for the company, but with minor impact in case of disclosure. Examples are configuration or calibration data
Negligible	No significant damage

Table 6.6 Image impact rating criteria for OEM or supplier

Impact rating	Criteria for image impact rating
Severe	News coverage in mainstream media, many people are concerned, including customers. Considerable investment required to improve reputation. Discontinuation of product possible. Loss of brand image
Major	News coverage in a circle of experts, such as technical magazines or academic conferences. Loss of customer confidence
Moderate	No news coverage, only few people concerned, but a possible public report of customers
Negligible	No significant damage

6.4.3.2 Image/Reputation

Cybersecurity attacks on major corporations are likely to end up in the mainstream media. In today's interconnected world, such news can be distributed around the world virtually instantly via the Internet. In the automotive industry, the biggest reputational damage will be to the automakers. At least to the public. Large suppliers can also suffer significant reputational damage, not from the end user buying the car, but from their own customers and car manufacturers. While attacks on vehicle components may not always end up in the mainstream media, they can still be presented at security conferences and shared among experts. We describe the different impact rating criteria for image damage in Table 6.6.

6.4.3.3 Financial

When analyzing the impact of cybersecurity attacks on companies such as OEMs and their suppliers, it is clear that the financial impact is secondary to the damage to human life. In a more general, abstract form, all other impact categories converge on financial damage. However, with the financial category for the TARA, we want to explicitly describe the direct impact of a cybersecurity issue. We opt for a generic and qualitative description. Each company can, if it wishes, assign a quantitative value, e.g., in millions of dollars, to the rest of the impact criteria. For example, IEC 62443 does this. The proposed financial impact evaluation criteria for OEM or supplier are described in Table 6.7.

Table 6.7 Financial impact rating criteria for OEM or supplier

Impact rating	Criteria for financial impact rating
Severe	Expensive compensations and eventual buyback of products
Major	Call-back of affected vehicles to patch affected software or replace parts
Moderate	The cybersecurity problem generated by the threat can be solved with a software patch
Negligible	No significant damage

6.4.3.4 Operational

In terms of a production environment, operational damage can be translated into a production line stoppage. A production outage following a cybersecurity issue can be considered a worst-case scenario. In Table 6.8 we describe different outage situations for the different impact rating levels. The values are based on IEC 62443.

6.4.4 Scalability

The term *scalability* in the context of damage scenarios is used to evaluate whether a damage scenario affects a single road user or multiple road users simultaneously. ISO/SAE 21434 does not provide specific examples of scalability, but allows scalability to be included in organization-specific rating criteria.

For example, EVITA [50] includes the concept of scalability (without using this term) in the description of the impact criteria for the categories. We present EVITA's classification in Table 6.9. The content related to scalability, in this case when multiple vehicles are affected, is present at severity levels two and three. We have highlighted these in bold. Note that the text describing the impact on multiple vehicles corresponds to the severity class below. When performing the severity rating, if a threat

Table 6.8 Operational impact rating criteria for OEM or supplier

Impact rating	Criteria for operational impact rating
Severe	More than 7 days manufacturing outage at one site, or more than 1 day manufacturing outage at multiple sites
Major	More than 2 days manufacturing outage at one site, or more than 1 h manufacturing outage at multiple sites
Moderate	Less than 1 day manufacturing outage at one site, or less than 1 h manufacturing outage at multiple sites
Negligible	No significant damage

Table 6.9 Severity classification extracted from *Table 4* of EVITA D2.3 [50]

Security threat severity class	Aspects of security threats			
	Safety	**Privacy**	**Financial**	**Operational**
0	No injuries	No unauthorized access to data	No financial loss	No impact on operational performance
1	Light or moderate injuries	Anonymous data only (no specific driver of vehicle data)	Low-level loss (~10)	Impact not discernible to driver
2	Severe injuries (survival probable) **Light/moderate injuries for multiple vehicles**	Identification of vehicle or driver **Anonymous data for multiple vehicles**	Moderate loss (~100) **Low losses for multiple vehicles**	Driver aware of performance degradation **Indiscernible impacts for multiple vehicles**
3	Life threatening (survival uncertain) or fatal injuries **Severe injuries for multiple vehicles**	Driver or vehicle tracking **Identification of driver or vehicle, for multiple vehicles**	Heavy loss (~1000) **Moderate losses for multiple vehicles**	Significant impact on performance **Noticeable impact for multiple vehicles**
4	Life threatening or fatal injuries for multiple vehicles	Driver or vehicle tracking for multiple vehicles	Heavy losses for multiple vehicles	Significant impact for multiple vehicles

affects multiple vehicles, it is rated according to the next severity class. For example, if the threat has the privacy rate *Anonymous data only*, it is assigned a severity of one. However, if the threat had the privacy rate *Anonymous data only*, but affected multiple vehicles instead of just one, then the threat would have a severity of two instead of one.

Despite the above examples from EVITA, to the best of our knowledge there is no general consensus nor known practical approaches for applying scalability to risk assessment for the automotive industry. For this reason, we do not pursue this concept further. However, scalability can be applied when modeling threats for IT/OT analysis. We have shown in Table 6.8 how the operational damage (based on IEC 62443) for a manufacturing plant can be affected by the scale of the attack (e.g., when only one site is affected or when multiple sites are affected). Applying scalability when analyzing the road user, for example, may raise some discussion whether it is acceptable or even reasonable to assess the impact when one or more road users are affected. For corporate damage, for example, it might be reasonable to consider the impact on reputation. While a remote attack on a single vehicle is unlikely to be noticed beyond the local media, a fleet-wide attack can be quickly publicized in the global media. On the other hand, the impact on the road user may not be directly affected by scalability. For example, human life is the most important entity to protect. In this case, it might be arguable to change the safety impact criteria in light of the attacker's scalability.

6.5 Practical Case Studies

We continue with the same case studies described in Chap. 5. We already defined the scope of each case study in Chap. 5, and in this section we identify the relevant assets and damage scenarios for each case study. The lists of assets and damage scenarios are not exhaustive due to space limitations. Rather, we provide real-world examples of the most common assets and damage scenarios.

6.5.1 Case Study 1: Automotive Embedded Device

In Sect. 5.7.1 we defined the scope of our TARA case study, including a description of the operational environment of the head unit ECU. Based on the analysis of the head unit interfaces in combination with the operational environment, several assets can be identified. As shown in Fig. 6.2, we follow the interactive steps of the asset identification process. We cannot simulate all the steps in this book, but let us assume that the process involves a collaborative analysis with various experts who know the details of the head unit. The asset identification process also involves considering the possible threats and damage scenarios that could occur. For example, our head unit ECU has high connectivity capabilities, such as direct access to the Internet, and exchanges data with the rest of the vehicle. It also has diagnostic interfaces. A starting point for asset identification could be to look at what could be the negative consequences for the road user if someone manages to introduce manipulated software via a diagnostic interface. Or even worse, what if manipulated software updates could be introduced over the Internet? From these two trivial questions, some assets can already be identified, such as the software update (as a software package or as an update system) or the diagnostic interfaces. The list of assets and their relevant cybersecurity properties are shown in Table 6.10.

Together with the identification of the relevant assets, the damage scenarios for our TARA case study need to be identified and described at this point. In Table 6.11 we present a list of exemplary damage scenarios for the road user. Some damage scenarios may have already been identified during the consideration of the assets and possible cybersecurity threats. For the sake of completeness, a practical approach to deriving the damage scenarios might be as follows: for each asset identified in Table 6.10, we analyze which damage scenarios can be identified if the cybersecurity properties of the asset are somehow compromised. At this point, it is not clear,

Table 6.10 List of assets and cybersecurity properties for our Case Study 1

Asset [WP-15-02]	Comment/Relevant cybersecurity property for the asset and rationale of the relevance
CAN for Diagnostics	Suppose a CAN bus is used only for diagnostic purposes in the workshop. Software updates are possible via this bus **Availability:** The services must be available when needed, especially for diagnostic purposes **Authenticity:** Because of powerful diagnostic functions such as software updates, the tool initiating a communication must prove its authenticity
JTAG	**Authenticity:** Similar to CAN diagnostics, a JTAG interface provides powerful debugging capabilities. Therefore, access to JTAG must be authentic
USB	**Authenticity:** The head unit may allow some convenience functions over a USB interface, such as transferring music and video from an external device. Software updates via USB are also possible. In this case, the head unit ECU must verify the authenticity of the tool before allowing write access
Wi-Fi/Cellular/Bluetooth	These interfaces provide Internet connectivity and allow comfort functions for passengers inside the vehicle **Confidentiality:** It is important due to the open nature of RF signals. In addition, sensitive data such as PII or software updates are expected to be transmitted over these interfaces **Integrity:** Data transmission must not be tampered with, or at least tampering must be detectable **Availability:** Access to the Internet and back-end services is a core functionality of a modern head unit. The availability of wireless connectivity must be guaranteed **Authenticity:** While authenticity may be required at different communication layers, it is an essential property that shares dependencies with other properties such as confidentiality
Ethernet	The Ethernet interface provides services from the head unit ECU to the rest of the vehicle. For example, software updates or diagnostic services are transferred over Ethernet **Confidentiality:** Some services require data confidentiality protection **Integrity:** Data integrity must be guaranteed, at least for critical services such as software updates **Availability:** The interface provides important functionality to the rest of the vehicle, such as position via GNSS, and therefore high availability must be guaranteed **Authenticity:** Similar to confidentiality and integrity, critical services require authenticity

(Continued)

Table 6.10 (Continued)

Asset [WP-15-02]	Comment/Relevant cybersecurity property for the asset and rationale of the relevance
Application Software	Software used in the SoC for video processing and in the SoC for connectivity functions **Integrity:** The integrity of software is critical to prevent several types of threats **Availability:** Software provides the core functionality of the head unit, so availability is critical **Authenticity:** Not only is integrity protection critical, but the ECU must guarantee that the software it is running is authentic
Software Updates	**Integrity:** It must not be possible to modify a software update **Authenticity:** It is essential that software updates come from the true creator of the update, i.e., the OEM's or supplier's software department **Non-Repudiation:** Software updates make significant changes to the functions of the Head Unit, such as different handling of PII. Therefore, it must be verifiable that a software update was performed with the user's consent and who initiated the update **Accountability:** The provider of the software update must share the liability in case of malfunction of the new software
Secret Keys	**Confidentiality:** Secret keys, especially private and symmetric keys, must be confidential by definition **Integrity:** The head unit ECU relies on keys for various purposes, such as verifying software authenticity and encrypting communications. Therefore, integrity is essential for proper operation **Availability:** Similar to integrity, key availability is essential for the correct operation of various functions **Authenticity:** The authenticity of keys must be ensured, especially if key updates are planned
Configuration Parameters	Assuming configuration parameters, such as variant coding, performed during the initial setup of the ECU **Integrity:** The configuration provided by the OEM or supplier shall not be modified to prevent fraud
Camera Data	The head unit processes video data from various cameras around the vehicle **Confidentiality:** Video sequences may contain PII or other sensitive information and must be protected from loss of confidentiality **Integrity:** The integrity of video data is important to prevent various types of misuse, such as deleting or altering parts of video sequences **Availability:** Services that rely on cameras require full availability **Authenticity:** Authenticity can be related to ensuring that the data has been processed by the head unit, for example

Table 6.11 List of damage scenarios and impact rating evaluation, concerning the road user view, for our Case Study 1

Damage Scenario [WP-15-01]	Impact Rating [WP-15-04]							
	Road User View							
	Safety	**Comment**	**Financial**	**Comment**	**Operational**	**Comment**	**Privacy**	**Comment**
Vehicle or component (e.g., ECU) gets damaged	Negligible	No safety impact	Moderate	Repair costs, vehicle not available	Severe	By definition, vehicle not available	Negligible	No privacy leakage
Sensors/warning systems provide wrong information	Major	Wrong information might lead to accidents	Major	Worst-case accident. Reparations. Lawyer costs	Major	Significant annoyance of the driver	Negligible	Privacy not affected
Disclosure of personal data	Negligible	No safety impact	Major	Eventual fraud and lawyer costs	Moderate	Reparation or software update	Major	By definition
Unavailability of non-safety information (e.g., traffic information, music, address book)	Negligible	No safety impact	Moderate	Repair costs, vehicle not available	Severe	Information not available as it was expected	Negligible	Privacy not affected
Publication of vulnerability	Negligible	No impact	Negligible	No impact	Negligible	No impact	Negligible	No impact

Table 6.12 List of damage scenarios and impact rating evaluation, concerning the OEM view, for our Case Study 1

Damage Scenario [WP-15-01]	Impact Rating [WP-15-04]							
	OEM View							
	Intellectual Property	**Comment**	**Image/Reputation**	**Comment**	**Financial**	**Comment**	**Operational**	**Comment**
Vehicle or component (e.g., ECU) gets damaged	Negligible	No impact	Moderate	No serious image impact	Moderate	Software patch	Negligible	Production not affected
Sensors/warning systems provide wrong information	Negligible	No impact	Major	Present at public report	Moderate	Restitution in some cases	Negligible	Production not affected
Disclosure of personal data	Negligible	No impact	Severe	Issue has to be reported. It gets public	Major	High damage if several vehicles affected	Moderate	Short stop possible for updating software
Unavailability of non-safety information (e.g., traffic information, music, address book)	Negligible	No impact	Moderate	No serious image impact	Moderate	Software patch	Negligible	Production not affected
Publication of vulnerability	Moderate	Code might get reverse engineered	Major	Present at public report/conference	Moderate	Software patch	Moderate	Short stop possible for updating software

nor is it important, to know exactly which threats and attack paths are actually possible. The essential question is: what damage can occur to the road user if the cybersecurity property of the asset is somehow compromised? The damage scenarios already identified through asset identification, as well as damage scenarios from catalogs and older TARA projects, can certainly be analyzed and included in the current TARA.

In addition to the damage scenarios for the road user, we are also interested in analyzing the negative effects of cybersecurity threats for the OEMs that produce vehicles with the ECU in scope. For this reason, we present a list of damage scenarios for the OEM in Table 6.12. We emphasize that the analysis of damage scenarios other than for the road user is not part of the ISO/SAE 21434 standard. However, as is the case in many real-world projects, it is important to know the risks to the OEM or supplier from a business perspective. Harmonizing the adverse cybersecurity impact on a road user, while allowing TARA according to ISO/SAE 21434, with the adverse impact on the business aims to improve overall cybersecurity governance. In both cases, the evaluation of the impact assessment was done following the criteria rationale described along Sect. 6.4.2. The damage scenarios heard in this section are real-world examples. However, due to space limitations, the list is not exhaustive. A more complete list of damage scenarios can be found in Appendix A.2.

6.5.2 Case Study 2: Vehicle-Backend Architecture

The TARA scope of our second case study focuses on the interaction of a connected vehicle with its ecosystem. We follow the scope definition described in Sect. 5.7.2. This case study aims at analyzing adversarial situations to the entire vehicle backend system, but especially threats to data. The asset identification follows the same generic steps as shown in Fig. 6.2. A set of assets in the form of services and data can be easily derived from the operational environment. For example, services for remote diagnostics and entertainment. The services rely on sensitive data stored in the vehicle, such as payment information. We identify and list the assets, including their cybersecurity properties, in Table 6.13.

How Deep does Asset Identification go? It is important to note that the assets in this case study are at the system level. In this case, for example, we do not consider the internal elements of the vehicle to be assets. Sensitive data is an asset, and it is presented in two forms: 1) sensitive data stored in the vehicle, and 2) sensitive data in transit from the vehicle to the backend and vice versa. The goal of the TARA in this example is to provide system goals for the vehicle. For the sensitive data example, we specify in the asset identification that this sensitive data must be stored protected from unauthorized manipulation. In the next step of the TARA, the threat scenario mapping, there will certainly be threats with attack paths that originate from the manipulation of an internal element of the vehicle. The result of the analysis of this case study may be, for example, a cybersecurity goal that requires sensitive data to

Table 6.13 List of assets and cybersecurity properties for our Case Study 2

Asset [WP-15-02]	Comment/Relevant cybersecurity property for the asset and rationale of the relevance
Diagnostic Tools/Data	**Integrity:** Manipulation of diagnostic messages can have negative consequences, especially for write operations. Manipulation should not be possible, or at least should be detected **Authenticity:** The ECU must be sure that the diagnostic tool trying to communicate with it is an authorized entity
Remote Diagnostics	**Confidentiality:** Diagnostics messages may contain PII or technical data that could be of interest to an outsider for a variety of reasons. However, unauthorized access to the data, especially during transmission, must be prevented **Integrity:** Manipulation of diagnostic messages can have adverse consequences, especially for write operations. Manipulation must be impossible, or at least detectable **Availability:** Remote diagnostics shall be available from the system as an operational requirement **Authenticity:** Due to its remote nature, the ECU shall allow diagnostic services only to trusted operators
Location Services	**Integrity:** The ECU provides accurate location services and should not be tampered with. Location can come from multiple sources, and tampering shall be detected **Availability:** Location services shall be available to applications that rely on them. Availability can be ensured, for example, by having multiple sources for determining the position of the vehicle
OEM Cloud Services	**Confidentiality:** It is good practice to ensure end-to-end confidentiality when sensitive data is transmitted. This is especially important for wireless communications **Integrity:** Transmitted data must be protected from unauthorized manipulation **Availability:** Even if the OEM cloud services are not critical to the operation of the head unit or the vehicle, the user expects full availability of convenience services **Authenticity:** Essential property to avoid phishing or threats related to addressing malicious servers

(Continued)

Table 6.13 (Continued)

Asset [WP-15-02]	Comment/Relevant cybersecurity property for the asset and rationale of the relevance
	Non-repudiation: In our example, a necessary property for legal or commercial reasons (e.g., if the user can activate features remotely) **Accountability:** similar to non-repudiation, assuming a necessary property in our example due to legal or commercial issues (e.g., if the user can activate features remotely)
Third-party Cloud Services	**Confidentiality:** It is good practice to ensure end-to-end confidentiality when sensitive data is transmitted. This is particularly important for wireless communications **Integrity:** Transmitted data must be protected from unauthorized manipulation **Availability:** Even if the third-party cloud services are not critical to the operation of the head unit or the vehicle, the user expects full availability of comfort services **Authenticity:** Essential feature to avoid phishing or threats related to addressing malicious servers **Non-repudiation:** In our example, a necessary property for legal or commercial reasons (e.g., if the user can activate features remotely) **Accountability:** Similar to non-repudiation, assuming a necessary property in our example due to legal or commercial issues (e.g., if the user can activate features remotely)
Sensitive Data stored in the Vehicle	**Confidentiality:** Sensitive data must be protected from unauthorized access. This is usually achieved by encrypting the data, so that in case of physical access to the storage device and memory dumplings, the data remains secret **Integrity:** Integrity protection can be applied to any type of data. Sensitive data can be used for critical operations and therefore must be protected against manipulation **Authenticity:** Depending on the data, it may be necessary to prove the authenticity of the data and its owner
Sensitive Data in Transit	**Confidentiality:** Sensitive data must also be protected from unauthorized access while in transit **Integrity:** It must be possible to detect manipulation of data in transit. This is especially important for data that is critical to operations, or for commercial or personal use **Authenticity:** Depending on the data, it may be necessary to prove the authenticity of the data and its owner

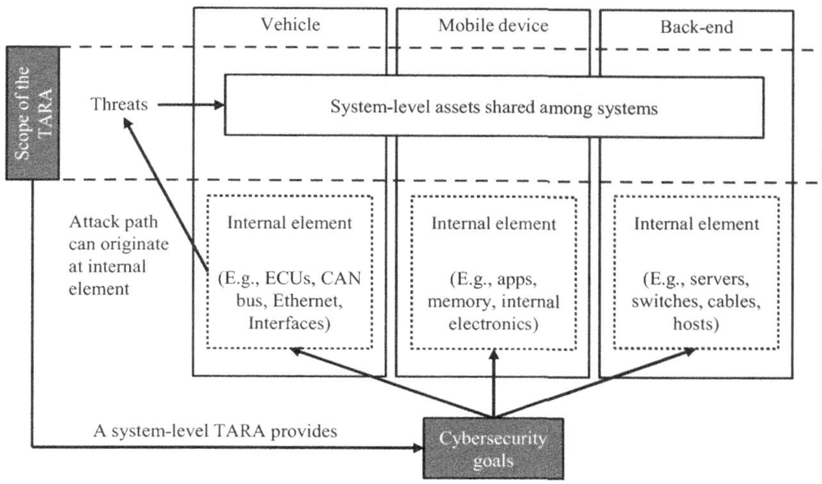

Fig. 6.4 Relationship between system-level assets, elements of the system, and cybersecurity goals

be protected when stored in the vehicle. The internal elements of the vehicle, such as an ECU or a CAN bus, are not assets for the scope of this case study. The goal of a TARA at this level is to establish high-level cybersecurity goals for an entire system. This is useful for complex scenarios, such as in this example, where different systems from different vendors share the same assets, such as *data*. System-level goals can be transferred to the entities responsible for the different elements of the system, such as back-end vendors, cell phone architects, and car manufacturers. A lower-level, or element-level, TARA can be performed for each of the individual systems, taking into account the previously provided system-level cybersecurity goals. The relationship of the system-level cybersecurity goals to the elements and scope of the TARA is shown in Fig. 6.4.

In this case study, we analyze the damage scenarios for the damage caused to the road user, as required by ISO/SAE 21434. In addition, the TARA case study will provide the adverse consequences of cybersecurity risks, from the same damage scenarios, for the OEM and for the third-party provider. Our goal is to show how a TARA, primarily used to satisfy ISO/SAE 21434, can be extremely useful for analyzing other damages and areas of an organization. In Table 6.14 we show the damage scenarios and their impact rating for the road user. We analyze the impact rating for the OEM view in Table 6.15 and for the third-party provider in Table 6.16. The third-party provider is used for online services that can be installed in the vehicle, such as a music or audiobook application.

A brief description of our damage scenarios and an example of a possible attack path is given below:

Table 6.14 List of damage scenarios and impact rating evaluation, concerning the road user view, for our Case Study 2

Damage Scenario [WP-15-01]	Impact Rating [WP-15-04]							
	Road User View							
	Safety	**Comment**	**Financial**	**Comment**	**Operational**	**Comment**	**Privacy**	**Comment**
Vehicle leaves trajectory or limited maneu-verability	Severe	Risk of collision/ crashing	Major	High costs due to infrastructure damage, and hospitalization	Severe	Vehicle is not operable	Negligible	No impact
Disclosure of business relevant data	Negligible	No impact	Negligible	No impact	Negligible	No impact	Negligible	No impact
Using services without valid license	Negligible	No impact	Moderate	Fines, or paying too much for the service	Negligible	No impact	Negligible	No impact
Remote services not available	Negligible	No impact	Moderate	Reparations, and time	Major	Services not working as expected	Negligible	No impact
Unauthorized software/data installed in the vehicle	Major	SW might overtake safety-functions of the vehicle	Moderate	Reparations, and time	Major	Operation of SW compromised	Negligible	No impact

Table 6.15 List of damage scenarios and impact rating evaluation, concerning the OEM view, for our Case Study 2

Damage Scenario [WP-15-01]	Impact Rating [WP-15-04]							
	OEM View							
	Intellectual Property	Comment	Image/ Reputation	Comment	Financial	Comment	Operational	Comment
Vehicle leaves trajectory or limited maneu-verability	Negligible	No impact	Severe	Accident published in the media world-wide	Severe	Restitutions, call-backs	Negligible	No impact
Disclosure of business relevant data	Major	Important data for competitors; engineering details	Major	High costs due to updates, and controls	Negligible	No impact	Negligible	No impact
Using services without valid license	Negligible	No impact	Negligible	No impact, or not relevant	Moderate	Restitutions to user, lawyer costs	Negligible	No impact
Remote services not available	Negligible	No impact	Moderate	Unsatisfied customers	Moderate	Restitutions, and patches	Negligible	No impact
Unauthorized software/data installed in the vehicle	Negligible	No impact	Major	Attack can get public at experts conferences	Major	Call-backs for updates, and engineering effort	Negligible	No impact

Table 6.16 List of damage scenarios and impact rating evaluation, concerning the third-party provider view, for our Case Study 2

Damage Scenario [WP-15-01]	Impact Rating [WP-15-04]							
	Third-party provider View							
	Intellectual Property	**Comment**	**Image/ Reputation**	**Comment**	**Financial**	**Comment**	**Operational**	**Comment**
Vehicle leaves trajectory or limited maneu-verability	Negligible	No impact	Negligible	No impact	Negligible	No impact	Negligible	No impact
Disclosure of business relevant data	Major	Provider data might be revealed	Moderate	Application patches might be needed	Negligible	No impact	Negligible	No impact
Using services without valid license	Negligible	No impact	Negligible	No impact	Negligible	No impact	Negligible	No impact
Remote services not available	Negligible	No impact	Moderate	Unsatisfied customers; Third-party provider's image affected	Moderate	Software patches	Negligible	No impact
Unauthorized software/data installed in the vehicle	Negligible	No impact	Negligible	No impact	Negligible	No impact	Negligible	No impact

- Vehicle leaves trajectory or limited maneuverability: The vehicle steers automatically, ignoring the steering wheel. An attacker defeats the vehicle's central gateway and gains access to the steering and braking ECUs. The attacker can send specially crafted CAN messages to these ECUs.
- Disclosure of business relevant data: Examples of business relevant data can be sales information, such as the number of vehicles produced in a given month. This information can be derived from IDs or counters stored in the vehicle during production. The mileage counter could also be used as an indicator.
- Using services without a valid license: One of the devices or applications connecting to external services has been compromised, and the attacker is able to gain access to services and functions without a valid license.
- Remote services not available: Services called directly from the head unit in the vehicle, or from other components of the vehicle, or from an external application, are partially or completely unavailable.
- Unauthorized software/data installed in the vehicle: The attacker gains access to write diagnostic functions of the vehicle and is able to install crafted software. The installation is either remote, through OEM and third party services, or local.

6.5.3 Case Study 3: Automotive Manufacturing Plant

In this third case study, we focus on manufacturing plants. The item of TARA is the manufacturing plants, including their machines and production lines. Similar to Case Study 2, this case study focuses on system-level components. The goal of the TARA is to identify risks in a manufacturing plant during normal operations and to provide cybersecurity goals for its components, such as internal servers, flash stations, or the ECU being produced. Therefore, the assets of this case study are at the component level. We consider the machines in the manufacturing plant, including their interfaces, but we do not further focus on the assets inside the machines. The scope definition is described in Sect. 5.7.3. The assets identified for this case study are summarized in Table 6.17.

The damage scenarios in this case study focus on damage to the road user, as required by ISO/SAE 21434, and to the supplier. We refer to the supplier, but only as an example. A similar analysis could be performed for the OEM. The list of damage scenarios is not exhaustive due to space limitations. The purpose of the case study is to provide practical examples of real damage scenarios to better understand the process. The selected damage scenarios affect the road user and the supplier at different levels. Some damage scenarios, such as the introduction of ransomware in production facilities, will, of course, not affect the road user in any way. However, it is one of the objectives of a TARA to consider such scenarios. The road user damage scenarios are mandatory as we want to comply with ISO/SAE 21434. The damage scenarios that only affect the production facility are relevant for cybersecurity governance. We describe the proposed damage scenarios as follows:

Table 6.17 List of assets and cybersecurity properties for our Case Study 3

Asset [WP-15-02]	Comment/Relevant cybersecurity property for the asset and rationale of the relevance
Debug/Diagnostic Interfaces at Flash Station, Production Line, and End of Line	**Integrity:** The integrity of the data exchange between a legitimate client (local or remote) and the diagnostic interface must be guaranteed to prevent or detect tampering
	Availability: Diagnostic services shall be reliable to prevent downtime in the event of failure
	Authenticity: Only trusted entities should have access to production machine diagnostic services
USB at Flash Station	Interface for manually updating the software of the flash station
	Integrity: Due to the criticality of the data transmitted, manipulation must be detectable
	Authenticity: Modification of the flash station software must only be done by authorized personnel. A tampered flash station can even introduce malware into the ECU being produced
Wi-Fi	**Confidentiality:** Due to the open nature of wireless signals, confidentiality is an essential property to prevent unauthorized access and various types of threats. Even if the data being transmitted is not sensitive per se, such as production monitoring data, it is necessary to protect the network packets
	Integrity: Protecting the integrity of Wi-Fi packets increases the robustness and reliability of the network and the systems that rely on it
	Availability: Wi-Fi networks used in production environments must be highly reliable
	Authenticity: Authentication of communication parties on a wireless network is essential due to its open nature
Fieldbus at All Stations	**Integrity:** Logistics or production data exchanged between machines must not be tampered with
	Availability: Production data exchanged between production machines is essential and must be highly reliable
Ethernet	**Integrity:** While quite dependent on the application layer, in general production data should not be tampered with in transit
	Availability: It is essential to production and must be available

(Continued)

Table 6.17 (Continued)

Asset [WP-15-02]	**Comment/Relevant cybersecurity property for the asset and rationale of the relevance**
	Authenticity: The authenticity of the data or the network devices connecting to the network must be ensured to prevent threats related to unauthorized access
Remote Access	**Confidentiality:** Especially relevant for data in transit. Remote access for diagnostic purposes is a powerful interface **Integrity:** Essential property for preventing or detecting tampering of diagnostic messages in transit **Availability:** Remote access can be used to provide rapid support in the event of production machine malfunctions. It is therefore important to ensure high availability **Authenticity:** It must not be possible for unauthorized entities to gain remote access to the production facility and production machines **Non-repudiation:** For critical services such as configuration changes or software updates to the machines, the entity using remote access must be traceable and uniquely identifiable **Accountability:** Similar to non-repudiation, traceability to the entity using the service must be ensured to provide accountability
Applications on Stations and Local Servers	**Integrity:** Software integrity must be ensured to prevent malware from running on production machines **Availability:** Essential for production **Authenticity:** Software installed on production machines shall be authentic. At least at the time of installation, it must be ensured that the software comes from a trusted source
Software Updates	**Integrity:** Protecting the integrity of software updates is essential to prevent malware from being installed on production machines **Authenticity:** Software update packages must come from trusted sources
Logistics Data	**Integrity:** Logistics data exchanged between production machines is essential **Availability:** Logistics data must be reliable because it is essential to production
Local Servers	**Integrity:** The software and hardware of local servers must not be tampered with

(Continued)

Table 6.17 (Continued)

Asset [WP-15-02]	Comment/Relevant cybersecurity property for the asset and rationale of the relevance
	Availability: Local services, such as those used to provide software to the ECU during production, or to store critical production information such as BOMs, must be highly available **Authenticity:** Important for the software running on the servers and for access to the network
Networking Devices	**Integrity:** The software and hardware of networking devices must not be tampered with **Availability:** Networking devices are essential to production and must be available **Authenticity:** Enforcing authenticity checks for devices connecting to the network is necessary to prevent threats associated with unauthorized access. In addition, access to the devices, such as for configuration purposes, must come from trusted entities
Security Artifacts from Key Management Servers	**Confidentiality:** Security artifacts, such as private keys for SecOC **Integrity:** Security artifacts are necessary for the proper functioning of the end product and must not be tampered with **Availability:** An essential part of the final product. Availability must be ensured **Authenticity:** Due to its use in cybersecurity, security artifacts must originate from trusted sources, such as the OEM's PKI

- Manipulated software introduced in the final product (ECU): The introduction of malware in the final produced ECU is probably one of the worst-case scenarios for both the manufacturer and the road user. The manufacturer is likely to suffer significant financial and reputational damage. Ransomware could be introduced in the ECU and executed a few years after production has started, leading to massive recalls. The road user is exposed to all kinds of consequences, such as loss of control of the vehicle and loss of privacy due to data breaches.
- Production machines and/or local servers infected with ransomware: Ransomware attacks are extremely damaging to production facilities, as they not only lead to a significant halt in production, but also cause privacy breaches. There have been several ransomware attacks that have compromised well-known automotive companies (see [51]).
- Production stop longer than one day: A longer production outage is the nightmare of every manufacturing company due to the financial losses involved.
- Large-scale data breach including cybersecurity artifacts and logistics data: Large-scale data breaches can affect the manufacturing company and the road user. In the worst case, PII, including customer and employee information, and cybersecurity artifacts are stolen from the manufacturing environment and used by attackers for extortion.

Table 6.18 List of damage scenarios and impact rating evaluation, concerning the road user view, for our Case Study 3

Damage Scenario [WP-15-01]	**Impact Rating [WP-15-04]**							
	Road User View							
	Safety	**Comment**	**Financial**	**Comment**	**Operational**	**Comment**	**Privacy**	**Comment**
Manipulated software introduced in the final product (ECU)	Severe	Assuming worst-case and the ECU takes over the operation of the vehicle leading to a crash with injuries	Major	Hospitalization costs; Reparations, and time lost	Severe	In worst-case scenario, the ECU and the vehicle do not operate as expected	Negligible	No impact
Production machines, and/or local servers infected with ransomware	Negligible	No impact	Negligible	No impact	Negligible	No impact	Negligible	No impact
Production stop longer than one day	Negligible	No impact	Negligible	No impact	Negligible	No impact	Negligible	No impact
Large-scale data breach including cybersecurity artifacts, and logistics data	Negligible	No impact	Negligible	No impact	Moderate	Loss of time for reparations	Negligible	No impact

(Continued)

Table 6.18 (Continued)

Damage Scenario [WP-15-01]	Impact Rating [WP-15-04]							
	Road User View							
	Safety	**Comment**	**Financial**	**Comment**	**Operational**	**Comment**	**Privacy**	**Comment**
Manipulated production data, or BOM, lead to defective final product	Severe	Worst-case scenario, vehicle crashes with injuries	Major	Hospitalization costs; Reparations, and time lost	Severe	ECU and vehicle do not operate as expected	Negligible	No impact

Table 6.19 List of damage scenarios and impact rating evaluation, concerning the supplier view, for our Case Study 3

Damage Scenario [WP-15-01]	Impact Rating [WP-15-04]							
	Supplier View							
	Intellectual Property	**Comment**	**Image/Reputation**	**Comment**	**Financial**	**Comment**	**Operational**	**Comment**
Manipulated software introduced in the final product (ECU)	Negligible	No impact	Major	Incident published in the automotive sector	Moderate	Due to software updates, and legal issues	Negligible	No impact
Production machines, and/or local servers infected with ransomware	Negligible	No impact	Severe	Coverage in mainstream media	Major	Due to production shutdown	Major	Assuming more than 2 days outage at one site
Production stop longer than one day	Negligible	No impact	Severe	Coverage in mainstream media	Major	Due to production shutdown	Major	Assuming more than 2 days outage at one site
Large-scale data breach including cybersecurity artifacts, and logistics data	Major	Assuming leakage of business-relevant data	Severe	Coverage in mainstream media	Major	Already produced ECUs to be updated, or replaced	Moderate	Shorter production stop according to BCP for recovering
Manipulated production data, or BOM, lead to defective final product	Negligible	No impact	Severe	Coverage in mainstream media	Severe	Call-backs due to defective hardware; Loss of business	Moderate	Shorter production stop according to BCP for recovering

- Manipulated production data, or BOM, lead to defective final product: A frustrated employee, or an attacker who get access to the internal network of the production plant, can manipulate production-relevant information, e.g., contained in databases, leading to the manufacturing of a defective production. If the manipulation is done carefully, e.g., by removing smaller components from a PCB, or just a bad parameterization of a component, the manipulation might not be detected during production (Tables 6.18 and 6.19).

References

1. ISO/SAE 21434:2021 Road vehicles—Cybersecurity engineering. Standard, International Organization for Standardization/Society of Automotive Engineers, Geneva, CH (2021). https://www.iso.org/standard/70918.html
2. European Union Agency for Cybersecurity (ENISA): ENISA good practices for security of Smart Cars (2019). https://www.enisa.europa.eu/publications/smart-cars
3. ISO 11898-1:2015 Road vehicles—Controller area network (CAN)—Part 1: Data link layer and physical signalling. Standard, International Organization for Standardization (2005). https://www.iso.org/standard/63648.html
4. ISO 11898-2:2016 Road vehicles—Controller area network (CAN)—Part 2: High-speed medium access unit. Standard, International Organization for Standardization (2016). https://www.iso.org/standard/67244.html
5. ISO 15765-2:2016 Road vehicles—Diagnostic communication over Controller Area Network (DoCAN)—Part 2: Transport protocol and network layer services. Standard, International Organization for Standardization (2016). https://www.iso.org/standard/66574.html
6. ISO 17458-1:2013 Road vehicles—FlexRay communications system—Part 1: General information and use case definition. Standard, International Organization for Standardization (2013). https://www.iso.org/standard/59804.html
7. Smith, C.: The Car Hacker's Handbook: A Guide for the Penetration Tester, 1 edn. No Starch Press, San Francisco (2016)
8. ISO 17987-1:2016 Road vehicles—Local Interconnect Network (LIN)—Part 1: General information and use case definition. Standard, International Organization for Standardization (2016). https://www.iso.org/standard/61222.html
9. Grzemba, A.: The MOST System. Franzis Verlag GmbH (2011)
10. ISO 13400-2:2019 Road vehicles—Diagnostic communication over Internet Protocol (DoIP)—Part 2: Transport protocol and network layer services. Standard, International Organization for Standardization (2019). https://www.iso.org/standard/74785.html
11. AUTOSAR FO requirements on SOME/IP protocol (2016). https://www.autosar.org/fileadmin/standards/R20-11/FO/AUTOSAR_RS_SOMEIPProtocol.pdf
12. ISO 14230-2:2016 Road vehicles—Diagnostic communication over K-Line (DoK-Line)—Part 2: Data link layer. Standard, International Organization for Standardization (2016). https://www.iso.org/standard/69115.html
13. ISO 9141-2:1994 Road vehicles—Diagnostic systems—Part 2: CARB requirements for interchange of digital information. Standard, International Organization for Standardization (1994). https://www.iso.org/standard/16738.html
14. Hug, N., Gibson, C., Kropotov, V., Vosseler, R.: Cybersecurity for Connected Cars (2021). https://documents.trendmicro.com/assets/white_papers/wp-cybersecurity-for-connected-cars-exploring-risks-in-5g-cloud-and-other-connected-technologies.pdf
15. Upstream: Upstream 2022 Global Automotive Cybersecurity Report (2022). https://upstream.auto/2022report/

16. Guidelines 1/2020 on processing personal data in the context of connected vehicles and mobility related applications (2020). https://edpb.europa.eu/our-work-tools/documents/public-consultations/2020/guidelines-12020-processing-personal-data_en
17. Rooting Bosch lcn2kai Headunit. https://github.com/ea/bosch_headunit_root
18. Regulation (EU) 2016/679 of the European parliament and of the council of 27 April 2016 on the protection of natural persons with regard to the processing of personal data and on the free movement of such data, and repealing directive 95/46/EC (general data protection regulation) (2016). https://eur-lex.europa.eu/eli/reg/2016/679/oj
19. tachomanipulation bei jedem dritten auto
20. United Nations: UN Regulation No. 155—Cyber security and cyber security management system E/ECE/TRANS/505/Rev.3/Add.154 (2021). https://unece.org/transport/documents/2021/03/standards/un-regulation-no-155-cyber-security-and-cyber-security
21. Regulation (EU) 2019/2144 of the European Parliament and of the Council of 27 November 2019 on type-approval requirements for motor vehicles and their trailers, and systems, components and separate technical units intended for such vehicles, as regards their general safety and the protection of vehicle occupants and vulnerable road users, amending Regulation (EU) 2018/858 of the European Parliament and of the Council and repealing Regulations (EC) No 78/2009, (EC) No 79/2009 and (EC) No 661/2009 of the European Parliament and of the Council and Commission Regulations (EC) No 631/2009, (EU) No 406/2010, (EU) No 672/2010, (EU) No 1003/2010, (EU) No 1005/2010, (EU) No 1008/2010, (EU) No 1009/2010, (EU) No 19/2011, (EU) No 109/2011, (EU) No 458/2011, (EU) No 65/2012, (EU) No 130/2012, (EU) No 347/2012, (EU) No 351/2012, (EU) No 1230/2012 and (EU) 2015/166 (text with EEA relevance) (2019). https://eur-lex.europa.eu/eli/reg/2019/2144/oj
22. Kreutner, M., Lauterwasser, C., Gwehenberger, J.: Investigating Accidents Involving Highly Automated Vehicles: Concept of a Data Trustee and Data Model for Future Homologation. p. 0035. National Highway Traffic Safety Administration (NHTSA) (2019). https://www-esv.nhtsa.dot.gov/Proceedings/26/26ESV-000035.pdf
23. Wurm, M.: Automotive Cybersecurity: Security-Bausteine für Automotive Embedded Systeme. Springer Berlin Heidelberg, Berlin, Heidelberg (2022). https://doi.org/10.1007/978-3-662-64228-3. https://link.springer.com/10.1007/978-3-662-64228-3
24. statista: Wertschöpfungsanteil der Automobilzulieferer am weltweiten Automobilbau in den Jahren 1985 bis 2015. https://de.statista.com/statistik/daten/studie/162996/umfrage/wertschoepfungsanteil-der-automobilzulieferer-am-automobilbau-weltweit/
25. United Nations: UN regulation no. 156—software update and software update management system E/ECE/TRANS/505/Rev.3/Add.155 (2021). https://unece.org/transport/documents/2021/03/standards/un-regulation-no-156-software-update-and-software-update
26. Roufa, I., Mustafaa, H., Taylora, T., Ohb, S., Xua, W., Gruteserb, M., Trappeb, W., Seskarb, I.: Security and privacy vulnerabilities of in-car wireless networks: A TirePressure monitoring system case study. pp. 11–13. USENIX Association (2010). https://www.usenix.org/legacy/events/sec10/tech/full_papers/Rouf.pdf
27. Jullien, F.: Renes'hack (2021). https://www.collshade.fr/articles/reneshack/rx_glitch_article.html
28. NIST: CVE-2021-43327 (2021). https://nvd.nist.gov/vuln/detail/CVE-2021-43327
29. IEC 61158-1:2019 Industrial communication networks—Fieldbus specifications—Part 1: Overview and guidance for the IEC 61158 and IEC 61784 series. Standard, International Electrotechnical Commission (2019). https://webstore.iec.ch/publication/59890
30. MODBUS: MODBUS Application Protocol Specification. Technical Report. https://modbus.org/docs/Modbus_Application_Protocol_V1_1b3.pdf
31. ODVA: Technology Overview Series: ControlNet Pub 200 (2016). https://www.odva.org/wp-content/uploads/2020/05/PUB00200R1-Tech-Series-ControlNet.pdf
32. PROFIBUS Nutzerorganisation e. V. (PNO)/PROFIBUS & PROFINET International (PI): PROFIBUS system description—technology and application (2016). https://www.profibus.com/download/profibus-technology-and-application-system-description
33. PROFIBUS Nutzerorganisation e. V. (PNO). https://www.profibus.com/technologies/profibus

34. Watson, V., Lou, X., Gao, Y.: A review of PROFIBUS protocol vulnerabilities—considerations for implementing authentication and authorization controls. In: Proceedings of the 14th International Joint Conference on e-Business and Telecommunications, pp. 444–449. SCITEPRESS—Science and Technology Publications (2017). https://doi.org/10.5220/0006426504440449. http://www.scitepress.org/DigitalLibrary/Link.aspx?doi=10.5220/0006426504440449
35. BSI industrial control system security: Top 10 threats and countermeasures 2022 (2022). https://www.allianz-fuer-cybersicherheit.de/SharedDocs/Downloads/Webs/ACS/DE/BSI-CS/BSI-CS_005E.html?nn=133710
36. IEC TR 62443-3-1:2009 Industrial communication networks—Network and system security—Part 3-1: Security technologies for industrial automation and control systems. Standard, International Electrotechnical Commission (2009). https://webstore.iec.ch/publication/7031
37. ISO 26262:2018 Road vehicles—Functional safety, Part 1–12. Standard, International Organization for Standardization (2018)
38. ISO 21448:2022 Road vehicles—Safety of the intended functionality. Standard, International Organization for Standardization (2022). https://www.iso.org/standard/77490.html
39. Lautenbach, A., Mafijul Islam, C.: HEAling vulnerabilities to ENhance software security and safety (HEAVENS)—deliverable d2—security models (2016). https://autosec.se/wp-content/uploads/2018/03/HEAVENS_D2_v2.0.pdf
40. IEC TS 62443-1-1:2009 Industrial communication networks—Network and system security—Part 1-1: Terminology, concepts and models. Standard, International Electrotechnical Commission (2009). https://webstore.iec.ch/publication/7029
41. ISO/IEC 27000:2018 information technology—security techniques—information security management systems—overview and vocabulary. Standard, International Organization for Standardization/International Electrotechnical Commission (2018). https://www.iso.org/standard/73906.html
42. Schneier, B.: Applied Cryptography: Protocols, Algorithms and Source Code in C, 20 edn. Wiley, Indianapolis, IN (2015)
43. Bosworth, S., Kabay, M.E.: Computer Security Handbook, 4 edn. Wiley, New York (2002)
44. NIST: NIST Special Publication 800-37 Risk Management Framework for Information Systems and Organizations—A System Life Cycle Approach for Security and Privacy (2018). https://doi.org/10.6028/NIST.SP.800-37r2
45. NIST: NIST Special Publication 800-53 Security and Privacy Controls for Information Systems and Organizations (2020). https://csrc.nist.gov/publications/detail/sp/800-53/rev-5/final
46. NIST: NIST Special Publication 800-12 An Introduction to Information Security (2017). https://csrc.nist.gov/publications/detail/sp/800-12/rev-1/final
47. ISO/IEC 29100:2024 Information technology—Security techniques—Privacy framework. Standard, International Society of Automation/International Electrotechnical Commission (2024). https://www.iso.org/standard/85938.html
48. IEC 62443-2-1:2010 Industrial communication networks—Network and system security—Part 2-1: Establishing an industrial automation and control system security program. Standard, International Electrotechnical Commission (2010). https://webstore.iec.ch/publication/7030
49. Gelinne, J.P.: The hidden costs of an IP breach: Cyber theft and the loss of intellectual property (2016). https://www2.deloitte.com/us/en/insights/deloitte-review/issue-19/loss-of-intellectual-property-ip-breach.html
50. Ruddle, A., Ward, D., Weyl, B., Idrees, S., Roudier, Y., Friedewald, M., Leimbach, T., Fuchs, A., Gürgens, S., Henniger, O., Rieke, R., Ritscher, M., Broberg, H., Apvrille, L., Pacalet, R., Pedroza, G.: Security requirements for automotive on-board networks based on dark-side scenarios (EVITA Deliverable 2.3) (2009). https://doi.org/10.5281/zenodo.1188418
51. Upstream: Upstream 2023 Global Automotive Cybersecurity Report (2023). https://upstream.auto/2023report/

Chapter 7
Modeling and Evaluating Threat Scenarios

Abstract Identifying threat scenarios and assigning them a probability of occurrence (feasibility) is the essence of a TARA. In this chapter, we address several key challenges in performing TARA. We describe how to actually perform threat identification and discuss the various popular threat modeling approaches. We also discuss which threat catalogs to use and how to use them. We look at the attack paths and discuss the technical depth required and how to perform the attack feasibility evaluation. We conclude the chapter with our three real-world case studies.

This step of the TARA process involves identifying all possible threats to the assets in the TARA. The mapping between threats and assets is modeled by the cybersecurity properties of an asset. The central question in this step is whether a threat has the potential to compromise any of the cybersecurity properties of an asset. The mapping of a threat to the affected cybersecurity properties of an asset is defined in ISO/SAE 21434 as a *threat scenario*. Threats come from a variety of sources: the expertise of security professionals, specialized Internet sites such as the OWASP® Top 10 [1], industry standards, and others. It is important to note that there is no single definitive source for threats, and a comprehensive approach involves considering inputs from multiple reliable sources. The description of a threat scenario is complemented with an *attack path*, which is the description of the exact steps or actions required by an attacker to execute a threat scenario. Attack paths are described in this step of the TARA for each threat scenario. Finally, the feasibility evaluation determines the effort and capabilities required to actually perform the attacks. The threat definition process is summarized in Fig. 7.1.

Necessary input for this chapter:

- Assets and their cybersecurity properties
- Attacker model (if used)
- Assumptions and known weaknesses

The outcome of this chapter is:

- Threat scenarios
- Attack paths (or attack trees)
- Feasibility evaluation

R. do Carmo and A. Schlensog, *Automotive Threat Analysis and Risk Assessment in Practice*, https://doi.org/10.1007/978-3-662-69614-9_7

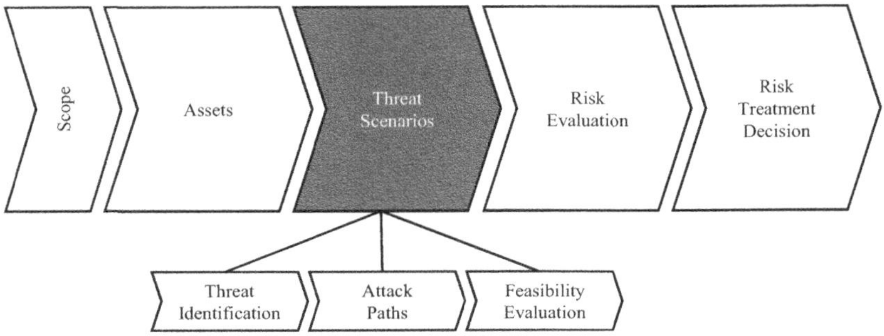

Fig. 7.1 High-level steps for the threat definition of the overall TARA process

7.1 Identifying Threats

Identifying the threat scenarios and their attack paths is the central point of creating a TARA. Threats are the reason assets are compromised and why there is a cybersecurity risk in the first place. Identifying threats is not only the most important step of the TARA, but also the most difficult and challenging. Finding threats is challenging for two reasons. First, there is no single approach to identifying threats. Threat identification is traditionally done by brainstorming, but there are also several structured approaches to threat modeling. Second, finding threats requires a good, deep technical knowledge of the assets, as well as cybersecurity techniques. It involves a creative process that puts the cybersecurity team performing TARA in the shoes of an attacker.

In the following sections, we briefly describe the most common threat modeling approaches. The list is not exhaustive, as our focus is on the automotive sector and in particular ISO/SAE 21434. There are four models that are explicitly mentioned in ISO/SAE 21434 as possible threat modeling approaches, namely EVITA, ETSI TVRA, PASTA, and STRIDE (see *[RQ-15-03]* in [2]). We take a special look at these models. There are more threat modeling approaches than those mentioned in this chapter, such as RACE [3], VAST [4], SAHARA [5], OCTAVE [6], LINDDUN [7], and others. A detailed overview of threat modeling for connected vehicles is available in [8]. We keep the description of the selected threat models short because we want to provide a practical insight into automotive TARA based on the international standard ISO/SAE 21434 and practical experience. There is plenty of literature on threat modeling that describes each model in detail, for example [9].

What is the Best Approach to Threat Identification? As explained by Shostack in [9] and in the Threat Modeling Manifesto [10], there is no single approach to threat modeling. Much like programming computer software, there is no ideal programming language. One programming language may be easier to write, while another may be more efficient with computer resources. In addition, it is recommended not to get caught up in strong opinions about what is a right or wrong threat model, but rather

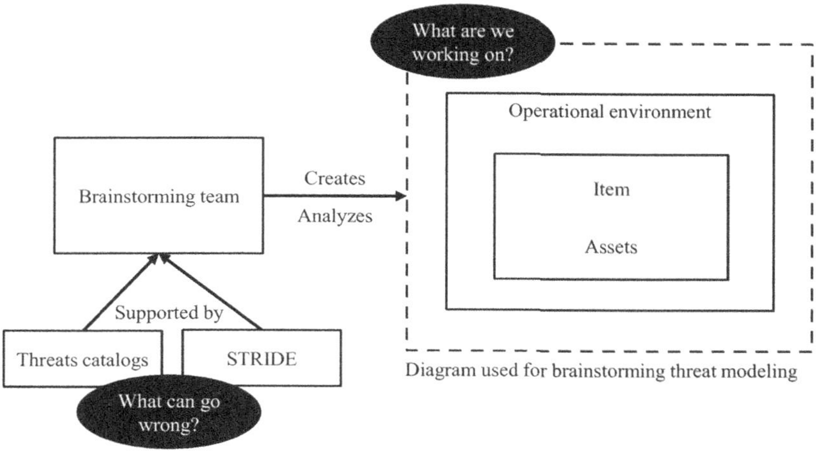

Fig. 7.2 Threat modeling process supported by brainstorming, threat catalogs, and STRIDE

to focus on clarifying the question of which threat model better fits the use case. Ultimately, the right threat model is the one that helps identify all possible threats in the most efficient way for a given project. In our experience, an efficient approach to threat modeling is brainstorming with the team performing TARA, supported by threat catalogs, and following STRIDE. Our threat modeling approach for TARA is shown in Fig. 7.2.

7.1.1 Brainstorming

Brainstorming is an effective and natural approach to threat modeling. The cybersecurity team, performing the TARA, sits down with experts in various fields related to the asset under analysis and brainstorms about threats. The team formed for the threat brainstorming session may be a team with the same stakeholders that were presented during the asset identification in Sect. 6.2. The brainstorming session can begin by analyzing a diagram of the asset and its operational environment. In principle, the threat modeling process should answer the following four key questions [9, 10]:

- *What are we working on?* In terms of ISO/SAE 21434, the important aspect behind this question is understanding the item and its operational environment. Understanding the item involves considering its major components and interfaces. It is important to analyze the boundaries of the item and how the item interacts with its environment. For example, when analyzing an automotive embedded device, the boundaries of the item might be the PCB and its interfaces to the outside world, such as CAN and Ethernet. Internal components of the item must also be considered, such as software components and memory areas. These components may already be assets of the TARA, as shown in Sect. 6.2.1. The use of diagrams may be the better approach for analyzing the item and its operational environment.

- *What can go wrong?* Once the diagram with the item and the operational environment is clear, it is time to analyze what can go wrong. In other words, identify all possible threats. At this point, we suggest using a systematic approach to analyze all the elements in the diagram. These elements are usually interfaces of the element and internal components. Since interfaces are usually the boundaries of an item, they form the so-called *attack surfaces*, i.e., the points on the boundary where an attacker can enter the system or cause some effect on it. The attack surface of the system may not be limited to interfaces. Internal components can also be the source of a threat, such as manipulated software performing some malicious action on the object after a certain time. Brainstorming only about threats runs the risk of forgetting about threats. Therefore, to ensure that no threat is forgotten and to give the brainstorming session a more structured approach, we recommend supporting this task with threat catalogs and STRIDE.
- *What are we going to do about it?* The question of how to deal with the threats and their respective risks is addressed by the TARA steps risk evaluation and risk treatment decision.
- *Did we do a good enough job?* When performing threat modeling through brainstorming, it is important to check the model for completeness. By definition, a model is a representation of a real system. As suggested in [9], the process of creating a diagram of the item and the operational environment can be done over several sessions. Each relevant stakeholder must verify that the model is complete and accurate.

Common attack vectors

While threat modeling depends on the particular system being analyzed, it is not a bad idea to consider the current state of common threat sources and attack vectors. A good summary of the most common attack vectors in recent years can be found in [11]. We list them as follows. The value in brackets represents the statistical frequency distribution.

- Servers (41.4%)
- Keyless entry / Key fob (26.3%)
- ECU/TCU/GW (12.2%)
- Mobile app (7.3%)
- Infotainment (5.7%)
- OBD port (5.4%)
- IT network (5.1%)
- Sensors (3.3%)
- Wi-Fi (2.9%)
- In-vehicle network (2.9%)
- Bluetooth (2.7%)

The list can at least be considered when creating diagrams and brainstorming threats. For example, questions related to the attack vectors above could be: can a compromised remote server attack my item?, what can happen if data transferred from the item to a cell phone is stolen?, or can a manipulated infotainment application compromise the vehicle bus?, among others.

7.1.2 Threat Catalogs

Threat catalogs are a useful tool to aid in the threat modeling process. The basic reason for having a catalog is that most TARAs are similar at some point. Obviously, the exact threat scenario is only applicable to a particular system because it involves exact steps in describing the attack path, and it is applied to a particular asset. However, many assets share many similarities across projects, and catalogs can be used as a first, rough approach at the beginning of the threat identification process. For example, interfaces such as CAN, Ethernet, Wi-Fi, or Bluetooth are common to many systems. With a pre-populated threat catalog, it is easier to select a large group of threats that are common to such interfaces. In addition, threat catalogs can have a lot of information pre-populated, such as typical cybersecurity properties affected, STRIDE classification, typical attack paths, and more.

The following sections describe the most common publicly available threat catalogs for automotive and IT/OT systems. However, none of the catalogs presented are intended to meet all the requirements of ISO/SAE 21434. They mostly provide a generic description of a threat, and in some cases a good technical description of how to execute the threat for some platforms. However, when transferring these threats to an ISO/SAE 21434 TARA, the cybersecurity team must still meet the requirements of the standard. For example, the affected cybersecurity properties of the asset and the detailed description of the attack path in the context of the associated asset must be described accordingly.

Do not only rely on catalogs

Catalogs are usually a convenient tool, and they can save a lot of time and increase productivity. As stated in [10], using toolkits to support threat modeling is a positive pattern. However, the downside of working with catalogs is that the cybersecurity team may be tempted to rely solely on them and lose rigor and creativity in brainstorming sessions. We strongly recommend using catalogs, but not relying solely on them.

7.1.2.1 UN Regulation No. 155 (Annex 5)

UN Regulation No. 155 [12] provides a list of threats and corresponding mitigations in its *Annex 5*. The threats focus on the vehicle itself, as well as back-end servers. The threats have a high-level description and an example of the vulnerability or attack method associated with the threat.

We strongly recommend evaluating the catalog of threats from UN Regulation No. 155. Evaluating these threats, at least by checking whether they are relevant to the item or not, and justifying the decision in writing, is mandatory for ensuring compliance with the regulation. Regardless of compliance, the list of threats presented in the regulation covers a wide range of threats and attack vectors that are quite interesting for modern connected vehicles. It includes attacks on back-end systems by insiders, including remote attacks, such as those enabled by a backdoor or an unpatched system, as well as local attacks. It also includes denial-of-service attacks on back-end servers, as well as recent major threats such as data breaches in the cloud. In-vehicle threats include manipulation and spoofing of communication channels, unauthorized access via the OBD port, and manipulation of internal CAN messages. Other in-vehicle threats include cryptographic key extraction, manipulation of diagnostic data, or extraction of vehicle owner PII. Compromising over-the-air software update procedures and software tampering are also listed.

7.1.2.2 MITRE ATT&CK®

MITRE ATT&CK® [13] is a very well-known, comprehensive framework and knowledge base of cybersecurity threats. The knowledge base is updated every two years, and the data comes from real-world attacks on publicly available threat intelligence and incident reporting sources. MITRE ATT&CK® was not originally designed for automotive embedded devices, but we highly recommend evaluating many of the threats presented there. Automotive embedded devices are becoming increasingly powerful and connected. For example, modern infotainment systems and head units already offer capabilities similar to cell phones. Many of them are already based on Android™ and iOS. In these cases, there is a direct correlation to the threats present at MITRE ATT&CK®. In the case of performing TARA for IT/OT systems, MITRE ATT&CK® provides threats for those very areas. In fact, the MITRE ATT&CK® knowledge base is organized into three broad areas called technology domains:

- *Enterprise:* Focuses on traditional networks and cloud technologies. The attacks are classified according to platforms Linux, macOS®, Microsoft Windows®, Amazon® Web Services, Microsoft® Azure, Google Cloud Platform™, Software-as-a-Service, Microsoft 365®, and Microsoft® Azure Active Directory.
- *Mobile:* Targets threats for mobile devices, for the platforms Android, and iOS.
- *ICS:* Targets threats for industrial control systems, e.g., threats that compromise control servers, field controllers, human-machine interface, or input/output servers.

Threats are modeled by tactics and techniques. Tactics represent the attacker's tactical objective, i.e., the reason for performing the attack. The high-level technical representation of how to achieve a tactic, i.e., the description of how to execute an attack, is modeled by techniques (and sub-techniques). Deeper technical insights into the techniques and sub-techniques are called procedures. The techniques, sub-techniques, and procedures from a threat of MITRE ATT&CK® can be useful for describing the attack path, when applying the threat to a TARA according to the ISO/SAE 21434. MITRE ATT&CK® also provides security concepts and classes of technologies to detect and prevent threats [13].

7.1.2.3 OWASP® Top 10

The OWASP® Foundation is a non-profit organization supported by an online community of experts dedicated to improving software security. They regularly publish the well-known report *OWASP® Top 10* [1] which describes the current ten most relevant cybersecurity risks for software developers and web application security today. The ten risks are well described, even practical examples are depicted. Preventive measures are also included. The cybersecurity risks listed are based on real-world observations, including data provided by users and surveys of industry professionals. The OWASP® team of experts consolidates all the data collected, including hundreds of CWEs. The CWEs associated with each risk are included in the risk description.

The vulnerabilities are primarily focused on web services and their backend infrastructure. However, we recommend considering these vulnerabilities for any type of system when conducting TARA. Almost all vulnerabilities, such as *insecure design*, *security misconfiguration*, and *cryptographic failures*, are generically applicable to any technology.

7.1.2.4 ENISA Threat Landscape (ETL)

Another reliable source of threats is the ENISA Threat Landscape (ETL) [14]. ENISA reports annually on the state of cybersecurity threats, including top threats and major trends in threats, threat actors (e.g., state-sponsored actors, cybercrime actors, hacktivists), and attack techniques. Each top threat is described in detail, with enough references and real-world data (e.g., from CWEs) to cover the motivation of the attackers behind each threat, what their target is, and what goals they are trying to achieve with the attack. The most relevant incidents within the analyzed time frame of one year are presented, including the organizations involved and the geographical spread of the incident. The ENISA Threat Landscape report includes a mapping between the top risks and the techniques of the MITRE ATT&CK® framework. The report provides recommendations for the most relevant threats. The recommendations are mapped to the measures provided by international standards, specifically ISO/IEC 27001 and the NIST Cybersecurity Framework (CSF).

7.1.2.5 ENISA Good Practices for Security of Smart Cars

A significant source of threats to smart cars has been published by ENISA in their report *ENISA good practices for security of Smart Cars* [15]. Smart cars are defined in the report as connected, autonomous and semi-autonomous vehicles. In this sense, the reported threats cover the entire autonomous vehicle ecosystem, from classic physical threats to the vehicle to modern threats to back-end systems and autonomous components. The report also provides a taxonomy of the most common assets found in intelligent vehicles. The assets are mapped to the threats provided in the report. In addition, cybersecurity measures and best practices are provided for each threat group.

7.1.2.6 Auto-ISAC Automotive Threat Matrix (ATM)

The Automotive Information Sharing and Analysis Center (Auto-ISAC) [16] is a recognized organization dedicated to fostering collaboration and information sharing in the area of automotive cybersecurity. Composed of automakers, suppliers, and other industry stakeholders, the Auto-ISAC serves as a trusted platform for sharing and analyzing cybersecurity threats, vulnerabilities, and best practices. Its members actively contribute by sharing emerging cybersecurity risks and discovered vulnerabilities related to vehicles, enabling a collective effort to effectively address and mitigate these potential risks.

Auto-ISAC published the Automotive Threat Matrix (ATM) [17], which is an enumeration of threats based on real-world observations. Additionally, ATM provides a common taxonomy for threats, including: Manipulate environment, Initial Access, Execution, Persistence, Privilege escalation, Defense evasion, Credential access, Discovery, Lateral movement, Collection, Command and control, Exfiltration, and Affect vehicle function. Some of the tactics (threats) are adapted from and mapped to MITRE ATT&CK®.

7.1.2.7 Automotive Security Research Group (ASRG)

The Automotive Security Research Group (ASRG) [18] is a non-profit, community-driven initiative to promote the development of security solutions for automotive products. Composed of leading experts in the field, the ASRG conducts research to identify vulnerabilities and develop effective security solutions for automotive systems. They also facilitate knowledge sharing and collaboration through conferences and workshops, advocate for robust cybersecurity practices, and actively contribute to industry standards. Community members can log in to their web portal and access a threat catalog. The threat catalog is updated by the community, and threats include a recommendation for feasibility rating and affected cybersecurity properties.

7.1.2.8 ISO/IEC 27000-Series

The ISO/IEC 27005 [19] part of the ISO/IEC 27000-series of international standards for ISMS also provides an informative list of sample threats and vulnerabilities (see *Annex A* in [19]). The threats and vulnerabilities presented in the standard do not have the level of technical detail found in other sources, such as MITRE ATT&CK®. However, they can be considered as a source of inspiration for a technical TARA. Depending on the requirements of the organization, it may be useful to consider these threats and vulnerabilities, or at least map them to other threats in the TARA, in order to have a similar baseline if the organization is already implementing an ISMS based on ISO/IEC 27001.

7.1.3 STRIDE

STRIDE is a widely used threat modeling approach developed by Loren Kohnfelder and Praerit Garg at Microsoft® [9, 20]. The acronym STRIDE is a mnemonic that stands for the first letter of the following threat categories:

- *Spoofing:* The attacker pretends to be someone else. In an automotive embedded system, this could take the form of a component or subsystem pretending to be another. For instance, an attacker could spoof an ECU sending false signals to the braking system. It might also include attempts to mimic communication between vehicle components or between the vehicle and external systems. Another example would be an attacker could potentially spoof signals from key fobs to unlock vehicles or start them remotely.
- *Tampering:* This refers to unauthorized modifications to data or system functionalities. In an automotive context, this could involve modifying the firmware of a vehicle component, altering sensor readings, or changing the software in an infotainment system to create a backdoor.
- *Repudiation:* The attacker's actions cannot be traced back to the attacker. For example, if the logging or auditing features of a vehicle's systems were compromised, it might be impossible to prove that a particular action (like a command sent to the vehicle's systems) happened.
- *Information disclosure:* The attacker gains unauthorized access to data, such as, but not limited to personal data (usually referred to as data breaches, or leaks).
- *Denial of service:* A situation where the attacker can disrupt the operation of a system or service. For example, a DoS attack could prevent the vehicle from functioning properly, such as by flooding the CAN bus with traffic, which prevents legitimate messages from getting through.
- *Elevation of privilege:* The attacker is able to gain more privileges than the attacker should. This might involve exploiting a vulnerability to gain access to a safety-critical ECU from a less-privileged system, like the infotainment system.

Table 7.1 STRIDE threats

Threat category	**Cybersecurity property violated**
Spoofing	Authenticity
Tampering	Integrity
Repudiation	Non-repudiation
Information disclosure	Confidentiality
Denial of service	Availability
Elevation of privilege	Authorization

The key idea behind STRIDE is that the cybersecurity team systematically considers the six threat categories during the threat modeling process. The STRIDE threat categories are the opposite of a number of important cybersecurity properties. The threat categories and violated cybersecurity properties are shown in Table 7.1.

STRIDE is a useful method for identifying threats because it addresses the fundamental question of what can go wrong. It is important to note that STRIDE is not a categorization of threats. STRIDE was not intended to be a threat taxonomy, even though it is commonly referred to as such. The goal of STRIDE is to help reason about attacks that can compromise the system within the scope of TARA. As explained in [9], using STRIDE to categorize threats is not helpful because the goal of STRIDE is to identify threats in the first place, e.g., during the brainstorming session. Worse, many threats do not fall into a single category, making the process of categorizing threats frustrating.

STRIDE-LM There is an extension of STRIDE that is often used in risk management, which is *lateral movement*. Particularly relevant to networks, this is the situation where the attacker gains control of the target network beyond the initial point of compromise. Lateral movement compromises the cybersecurity property of *containment* (or *segmentation*) [21].

7.1.4 EVITA

E-Safety Vehicle Intrusion Protected Applications (EVITA) is a pioneering project for cybersecurity risk assessment in the automotive industry [22]. Several relevant cybersecurity standards, such as SAE J3061 [23], and risk assessment approaches, such as HEAVENS, published afterward, are based on EVITA, or at least have used parts of the concepts of EVITA. The publication *Deliverable D2.3: Security requirements for automotive on-board networks based on dark-side scenarios* [24], published in 2009, provides security requirements based on risk analysis. EVITA's approach is based on the methodology of the international standard ISO/IEC 18045:2008 [25]. In addition, EVITA's hazard analysis and risk assessment methodology is based on the hazard analysis and risk assessment of the international functional safety standard ISO 26262[1] [26]. The threat impact classification used in EVITA is similar to that used today by the damage scenario impact rating evaluation in ISO/SAE 21434. EVITA has taken the classification used for functional safety in ISO 26262 and added three new categories: *privacy*, *financial*, and *operational*.

Threat identification is performed using *dark-side scenarios* and attack trees. The dark-side scenarios consist of the following components: identification and classification of possible attack motivations, evaluation of attacker capabilities, and attack modeling. The attack modeling step is supported by attack trees. The root of EVITA's attack trees represents the so-called attack goal, which is the illegal benefit to the attacker for compromising the identified use cases described in [27]. The second layer of the attack trees represents the attack objectives that could satisfy the attack goal. The attack objectives represent the harm to stakeholders (e.g., road users, service operators, OEMs and suppliers). The third layer of the attack trees represents the different possible attack methods that can be used to achieve an attack objective. The remaining layers represent the assets compromised and the intermediate steps. The attack potential is assigned to the asset attack entry points at the bottom of the tree.

EVITA's threat modeling methodology is well known and widely accepted in the automotive industry. It may be useful to consider using EVITA, especially by applying their attack tree approach and taking inspiration from the proposed threats. Another advantage of using EVITA is for backward compatibility reasons, if the organization already has an extensive database of TARAs created with EVITA. The obvious disadvantage of using EVITA is the mapping and compatibility issues when transferring the threat modeling results into an ISO/SAE 21434 TARA.

7.1.5 HEAVENS

Healing Vulnerabilities to Enhance Software Security and Safety (HEAVENS) is a risk assessment model designed by the automotive industry in 2016 [28]. The HEAVENS risk assessment model was designed to overcome some of the drawbacks of older models, particularly EVITA. For example, HEAVENS argues that EVITA is attack-centric and focuses on identifying all possible attacks against a target of evaluation. HEAVENS, on the other hand, is a threat-centric model supported by the STRIDE approach. In this way, HEAVENS focuses on a handful of generic threat categories rather than the many possible attacks and attack techniques associated with an asset. In addition, HEAVENS proposes a direct mapping between security attributes (e.g., confidentiality, integrity, availability) and threats. In addition to safety, financial, operational, and privacy, HEAVENS introduces legislation as a security objective.

The threat analysis of HEAVENS is based on STRIDE and extended to automotive E/E systems. Threats from other security models, such as EVITA, are analyzed

[1] Still at DIS stage at the time of EVITA.

by HEAVENS (cf. *table 4-1* in [28]) and mapped to STRIDE. HEAVENS has an interesting approach to distinguishing between threats and vulnerabilities, arguing that threats should be considered before vulnerabilities. If TARA is performed in the concept phase, it is easier to identify threats to the assets than vulnerabilities because implementation details may be missing. The introduction of vulnerabilities during product development and implementation is also considered by the model.

The creators of HEAVENS have proposed to create a new risk assessment model, called HEAVENS 2.0, which is based on HEAVENS, and it is fully compliant with the ISO/SAE 21434 in [29].

7.1.6 ETSI TVRA

We have already introduced the ETSI Threat, Vulnerability, and Risk Analysis (TVRA) model through our attacker modeling in Sect. 5.6.2. TVRA is a well-known and widely accepted risk assessment model. It was originally developed for the telecommunications industry, but is generally applicable to other domains. In fact, TVRA has been accepted in the automotive industry. HEAVENS analyzes TVRA in detail in [28], and ISO/SAE 21434 mentions TVRA as a possible framework for threat modeling.

For threat modeling, TVRA defines a set of threat groups that are relevant to telecommunications systems. These threats are *interception (eavesdropping)*, *unauthorized access*, *masquerade*, *forgery*, *loss or corruption of information*, *repudiation*, and *denial of service*. The cybersecurity properties evaluated by TVRA are confidentiality, integrity, availability, authenticity, and accountability. For the risk assessment process in TVRA, the first step is to identify weaknesses and vulnerabilities in the target of evaluation. Weaknesses are the potential point of attack for an attacker (or *threat agent* in terms of TVRA). If there are threats to weaknesses, and if the weaknesses are realistic to be attacked, they are considered vulnerabilities.

We recommend drafting threats according to TVRA if the organization needs to provide backward compatibility with existing TARAs. For new TARA projects, the benefits of using TVRA are limited. The threats are specific to telecommunications systems. They can be correlated to automotive systems, but we do not see a greater advantage with respect to the direct application of STRIDE. In addition, it requires additional effort from the TARA team to first understand TVRA and then apply it to automotive TARA.

7.1.7 PASTA

Process for Attack Simulation and Threat Analysis (PASTA) is a risk-centric, evidence-based threat modeling methodology with a particular focus on business processes. The goal of PASTA is to address threats that impact the business, financial, and operational objectives of the application being analyzed. For example, the very first steps of the PASTA methodology are to determine the business objectives for the application, service, or infrastructure under risk analysis and to identify the most critical business use cases. PASTA has become popular worldwide, especially in the business sector, but also in other major industries and organizations such as *GitLab* [30].

PASTA consists of seven stages: *1. Define Objectives*, *2. Define Technical Scope*, *3. Application Decomposition and Analysis*, *4. Threat Analysis*, *5. Weakness and Vulnerability Analysis*, *6. Attack Modeling and Simulation*, and *7. Risk Analysis and Management*. Threats are extracted from external sources such as OWASP. The threats are mapped to threat agents and attacks using attack trees [31]. The threats are given a probability value (i.e., a feasibility rating) based on considerations of access, windows of opportunity, ability to repudiate, risk reward for the threat agent, and simplicity of the threat [32]. A relevant phase of PASTA is the vulnerability analysis, which involves targeted vulnerability testing of the system under analysis. Once the vulnerability analysis is complete, the attack scenarios are consolidated and the risk is finally assessed.

7.2 Attack Paths

An attack path is the description of the exact steps or actions required by an attacker to execute a threat scenario. The identification of attack paths is called *attack path analysis* in ISO/SAE 21434. The attack paths are an essential part of the TARA because they describe how the threat scenarios can be realized for a given asset. The attack paths do not answer the question of what can happen, but how can it happen? There must be at least one attack path for each identified threat scenario. Otherwise, the threat scenario is not possible, at least in the context of the analyzed item or component.

7.2.1 How to Identify the Attack Paths?

Identifying attack paths is not a simple task and can take a considerable amount of time. The attack paths require technical knowledge of the item or component, as well as cybersecurity techniques, to provide accurate and realistic attack paths. If the attack paths are not accurate enough, it could lead to an incorrect risk calculation. For this reason, the best way to describe the attack paths is through a collaborative analysis between the item experts and the cybersecurity team. Similar to the threat identification process, brainstorming sessions can be very beneficial.

Attack paths can be identified using either a top-down or bottom-up approach. In the top-down approach, we start by analyzing a particular threat scenario and reasoning about how it can be realized. This is the typical case where attack trees become relevant. In the bottom-up approach, we start the analysis from identified vulnerabilities. Vulnerabilities can come from a vulnerability analysis as described in ISO/SAE 21434, from vulnerabilities found during product development, or from cybersecurity events. The vulnerabilities are analyzed and the possible path to creating threat scenarios is identified. The analysis of vulnerabilities may identify new threat scenarios. These new threat scenarios should be included in the TARA.

Detail level of the attack paths
There is usually a discussion about the level of technical details of the attack paths. There is no exact description of how to describe the attack paths. However, the whole purpose of defining attack paths is to analyze the probability of an attack actually happening. In effect, we are looking for risks. Risk, in its simplest definition, is the impact of something happening multiplied by the probability of it actually happening. The attack path should provide a description that is sufficient to allow an accurate estimate of the feasibility (i.e., probability) of an attack. In this sense, it may be that a description as short as *do fuzzing on UDS from DID 0x0 to DID 0xFFFF at CAN using the SID 0x2E* is sufficient to provide an accurate estimate of the feasibility of the attack.

7.2.2 Description of Attack Paths

The attack paths are described as a list of steps involved in the attack. The separation into steps is not only easy to read and understand, but it is also necessary to evaluate the feasibility of each step, as we discuss later in Sect. 7.3. Each step of the attack path is mandatory for the execution of the entire attack path. If there are steps that can be omitted or replaced by other steps, then different attack paths must be derived, specifying each of these variations.

For example, the following steps describe the attack path for a head unit ECU that implements the threat scenario *Unauthorized access to information on back-end servers*:

1. Rent a vehicle using falsified ID and credit card
2. Open vehicle and extract head unit ECU
3. Research how to get diagnostics access to the ECU
4. Get the necessary debugging tools and connect them to the ECU (e.g., using UDS, or unprotected JTAG)
5. Dump memory area where application code is stored and do reverse engineering on the code
6. Analyze which part of the code establishes the communication to the back-end server and sends requests to a database
7. Manipulate the binary so a simple SQL injection string can be added to the database query

Similar examples are available in the *Annex H, section H.2.5* of the ISO/SAE 21434 standard [2].

7.2.3 Attack Trees

Attack trees are a popular approach to modeling, identifying, and describing the variety of attack paths. Attack trees are nothing more than a data structure, but they are extremely useful to humans when presented in graphical form. In this way, they are a very practical approach when discussing ideas in a brainstorming session. Attack trees have been used extensively in the threat modeling community. One of the earliest descriptions of attack trees for information security was provided by Bruce Schneier in [33]. Schneier describes attack trees as a formal, methodical way of describing the security of systems based on varying attacks. In an attack tree, the goal of the attack is represented as the root node, while the different ways of achieving that goal are represented as leaf nodes [33]. The topic is discussed in detail in [9].

To identify attack paths, we can follow the structure of simple attack trees discussed in [33]. At the top of the tree, or root node, we define a threat scenario. The subsequent leaf nodes of the tree describe the steps involved in the attack paths that implement the threat scenario. Note that an attack tree can contain multiple attack paths that share similar steps. The combination of multiple attack paths is possible because the attack trees combine AND nodes and OR nodes. OR nodes represent alternatives to reach a leaf. For example, an attack could be *Get OTA data in transit* and have the child OR nodes *Capture cellular communication* and *Capture Wi-Fi communication*. AND nodes, on the other hand, represent the steps that are mandatory within an attack path to be realized. For example, to achieve the step *Capture Wi-Fi communication*, the attacker must perform both of the child steps *Be located near the vehicle* and *Turn on sniffer tool on laptop*.

We present an example of an attack tree that shows the attack path that we described in text form in Sect. 7.2.2 for a head unit ECU that implements the threat scenario *Unauthorized access to information on back-end servers* in Fig. 7.3.

Efficiency of Attack Trees While attack trees are very popular and useful, they have the potential drawback of wasting time dealing with issues related to graphical representation. Another problem we sometimes see is that the steps of an attack path must be transferred from the attack tree tool to another tool or spreadsheet where the TARA is performed. This is necessary, for example, to perform the feasibility evaluation of the steps of the attack paths. Of course, there are modern tools that combine attack trees and TARA at the same time. However, if this is not the case in the project, it should be considered whether it is possible to work with attack trees within the time frame allotted for the TARA.

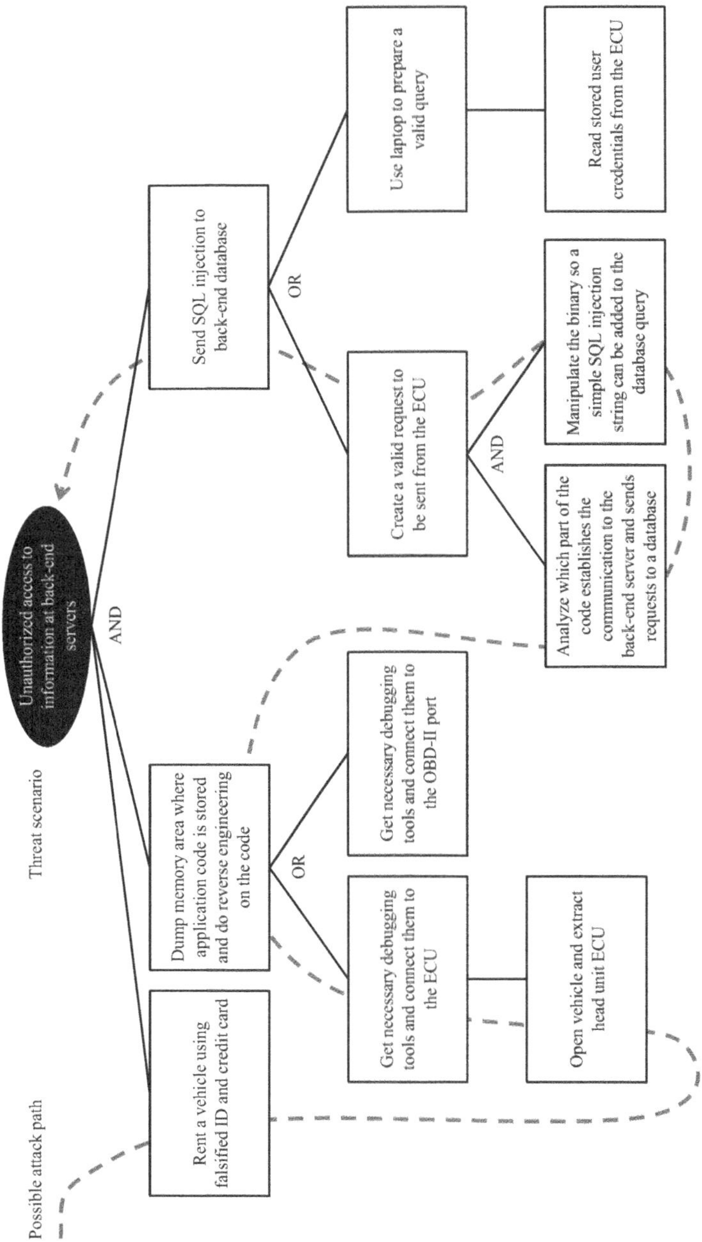

Fig. 7.3 Example of an attack three for possible attack paths for realizing a threat scenario at a head unit ECU

7.3 Feasibility Evaluation

The attack feasibility evaluation consists of determining the *attack feasibility rating*, which represents the effort required to create an attack, including all necessary steps and resources required to execute the attack. In terms of risk calculation, attack feasibility is the representation of the probability that an attack will occur. The attack feasibility rating is applied to each attack path that realizes a particular threat scenario. ISO/SAE 21434 defines four values for the attack feasibility rating, which we describe in Table 7.2. The lower the effort required to execute an attack path, the higher its attack feasibility rating.

ISO/SAE 21434 describes three possible approaches for determining the attack feasibility rating:

- *Attack potential-based approach*: Widely accepted approach for determining attack potential. A major advantage of this approach is its flexibility based on different categories and ratings. In addition, well-known models and frameworks in the cybersecurity and information security community, such as the Common Criteria [34], and in the automotive industry, such as ETSI TVRA [35], EVITA [24], and HEAVENS [28], use this method with the same or similar parameters.
- *CVSS-based approach*: CVSS is also a well-known framework, especially used for vulnerability rating. The attack feasibility is modeled by the exploitability parameter, which is calculated from rating categories, similar to the attack potential-based approach. The advantages of CVSS are that the categories may be easier to handle and the CVSS score can be used to compare vulnerabilities. However, the rating may not be as accurate as the attack potential-based approach because the number of categories is smaller, and, some ratings have only a few values to choose from. [36].
- *Attack vector-based approach*: This is the simplest approach to attack potential rating because it is based on only one parameter, the attack vector. The attack vector represents the context, basically remote or local, that the attacker uses to exploit the attack path. The attack vector is also present in the CVSS framework. As mentioned in ISO/SAE 21434, the attack vector-based approach may be appropriate in the early stages of product development because there is usually insufficient information (e.g., no implementation details, important architectural decisions are

Table 7.2 Attack feasibility ratings according to ISO/SAE 21434

Attack feasibility rating	Description
High	The attack path can be accomplished utilizing low effort.
Medium	The attack path can be accomplished utilizing medium effort.
Low	The attack path can be accomplished utilizing high effort.
Very low	The attack path can be accomplished utilizing very high effort.

missing, or final decisions on interfaces have not been made). An initial TARA performed using the attack vector-based approach can be reviewed at a later time by using the attack potential-based approach.

We describe them in detail in the following sections.

7.3.1 Attack Potential-Based Approach

The attack potential-based approach, based on the international standard ISO/IEC 18045 [25, 37], determines the attack feasibility rating based on five categories: *elapsed time*, *specialist expertise*, *knowledge of the item or component*, *window of opportunity*, and *equipment*. The rating categories presented in ISO/SAE 21434 are adaptations of ISO/IEC 18045. The rating values and their descriptions can be customized according to company-specific policies. Each category is described in this section.

About the inaccuracy of feasibility evaluation
The selection of the rating values of each category follows a theoretical approach in the vast majority of TARA projects, based on the subjective opinion of the evaluator. However, the selected rating should be as close to reality as possible. In fact, ISO/IEC 18045 states that penetration tests can be performed for each attack scenario to confirm or disprove the theoretical analysis (see *section B.6.2.4 Calculation of attack potential* in [37]). Of course, practical testing of each attack path would be the ideal case, but this is usually not practicable. For this reason, it is important that the team performing the rating has sufficient technical background. Relying on catalogs may help in the absence of sufficient practical experience, but each attack path is different from system to system, even if they describe a very similar exploit or action. Examples of attack feasibility ratings for an extensive list of attacks for various assets can be found in [36]. In the best-case scenario, an automotive penetration tester/team supports the activity by brainstorming or reviewing it afterward.

Each rating of the five categories is assigned a numerical value. The attack potential, as described in ISO/IEC 18045, is calculated by adding all values of the selected rating of each category. ISO/SAE 21434 proposes an adaptation of the values and parameters from ISO/IEC 18045 (cf. *table G.6* and *table G.7* in [2]). We show the values for the rating categories in Table 7.3, and the mapping of the resulting attack potential to the attack feasibility rating in Table 7.4.

Table 7.3 Calculation of the attack potential (*Table G.6* of ISO/SAE 21434 [2])

Elapsed time		Specialist expertise		Knowledge of the item or component		Window of opportunity		Equipment	
Rating	**Value**	**Rating**	**Value**	**Rating**	**Value**	**Rating**	**Value**	**Rating**	**Value**
≤ 1 day	0	Layman	0	Public	0	Unlimited	0	Standard	0
≤ 1 week	1	Proficient	3	Restricted	3	Easy	1	Specialized	4
≤ 1 month	4	Expert	6	Confidential	7	Moderate	4	Bespoke	7
≤ 6 months	17	Multiple experts	8	Strictly confidential	11	Difficult/ None	10	Multiple bespoke	9
> 6 months	19	–	–	–	–	–	–	–	-

Table 7.4 Mapping attack feasibility rating and attack potential

Attack feasibility rating	Attack potential values
High	0–9
	10–13
Medium	14–19
Low	20–24
Very low	≥25

7.3.1.1 Elapsed Time

The total amount of time it takes by an attacker to identify a vulnerability, develop an attack method (e.g., an exploit) and actually execute the attack, is represented by the category *elapsed time*. The ISO/SAE 21434 proposes an adaptation of the ISO/IEC 18045 rating scale. We illustrate the rating scale and provide practical examples as follows:

- *≤1 day:* The attack is simple, and the attacker already has enough knowledge and methods to carry it out. For example, a known vulnerability can be exploited using well-known tools available on the Internet. Another example can be an enumeration attack, i.e., by connecting a scanner (PC with CAN dongle) to the OBD-II port of a vehicle.
- *≤1 week:* An attacker with penetration testing experience can gain access to a locally available ECU that does not have sufficient cybersecurity controls. For example, a Python script can be used to brute force the bootloader password. Many of these examples can be found in [38].
- *≤1 month:* Less than a month can be the time it takes to build an exploit based on a known vulnerability. For example, the attacker may know that the item or component has a particular vulnerability, but a tool to exploit it, such as a script that sends and interprets CAN messages, must first be programmed.
- *≤6 months:* The attacker may not need much time, but for example, the attacker may want to compromise a specific vehicle.
- *>6 months:* Typical example of APT attacks, or attacks that require some planning and complex logistics. It may include research time spent to find a vulnerability on a specific system.

7.3.1.2 Specialist Expertise

The *expertise* refers to the level of knowledge the attacker must have in order to execute the attack. Knowledge can be general knowledge of the underlying principles, expertise in specific areas, or attack methods. We present the rating scale provided in ISO/SAE 21434 with examples as follows:

- *Layman:* Unknowledgeable compared to experts or skilled individuals. A typical example is the script kiddie, a person with little or no technical background who just wants to experiment with tools and step-by-step instructions that are easily available on the Internet. A layman can also be a car thief using standard tools to intercept or relay key fob signals.
- *Proficient:* Knowledgeable in that they are familiar with the security behavior of the product or system type. A knowledgeable person may be an experienced vehicle owner, a repair shop professional, or a tuner who can perform common attacks such as odometer tampering or the installation of counterfeit parts.
- *Expert:* Experts are familiar with the underlying algorithms, protocols, hardware, structures, security behavior, security principles, and engineering concepts used in a vehicle's design. They are also familiar with techniques and tools for defining new attacks, cryptography, classic attacks for the product type, attack methods, and vulnerabilities present in the product or system type. Examples of experts are engineers (e.g., employees or competitors in the automotive sector), motivated hackers, IT security professionals, or university researchers.
- *Multiple experts:* Different areas of expertise are required at the expert level for different steps of an attack. Examples include research institutions and laboratories, cybercriminal organizations, or government-sponsored organizations (intelligence services).

7.3.1.3 Knowledge of the Item or Component

Knowledge of the item or component is a metric for describing the specific knowledge of the system required to execute the attack. The evaluation according to ISO/SAE 21434 including examples is described as follows:

- *Public information:* Public information about the item or component (e.g., as obtained from the Internet). For example, public documentation about the product, Internet forum, books, or specifications from standards and protocols used (e.g., CAN, Ethernet, AUTOSAR).
- *Restricted information:* Restricted information about the item or component (e.g., knowledge controlled within the development organization and shared with other organizations under a non-disclosure agreement). Examples include internal documentation shared between manufacturer and supplier, such as requirements or design specifications.
- *Confidential information:* Confidential information about the item or component (e.g., knowledge that is shared among discrete teams within the development organization and access to which is restricted to members of the specified teams). Examples include immobilizer-related information, source code, or other sensitive design details. Examples also include cybersecurity-related documentation, such as the results of penetration tests or vulnerability analyzes (assuming they are not fixed, or the report contains details that could be used to find new vulnerabilities or easily compromise the system).

- *Strictly confidential information:* Strictly confidential information about the item or component (e.g., knowledge that is known by only a few individuals, access to which is very tightly controlled on a strict need to know basis and individual undertaking). Examples include custom calibrations or memory maps that are internally documented by the manufacturer or supplier.

7.3.1.4 Window of Opportunity

The *window of opportunity* determines the amount of access to the item or component needed to perform the attack. The window of opportunity combines the type of access (e.g., local and physical) and the duration of access (e.g., unlimited and limited). The evaluation according to ISO/SAE 21434, including examples, is described as follows:

- *Unlimited:* Highly available over public/untrusted network without any time limitation (i.e., asset is always accessible). Remote access without physical presence or time limit and unlimited physical access to the item or component. Examples include remote attack without any preconditions and unlimited physical access by the owner for chip tuning.
- *Easy:* Highly available and time-limited access. Remote access without the physical presence of the item or component. Examples include Bluetooth pairing time, remote software updates, and remote attacks that require the vehicle to be stationary.
- *Moderate:* Low availability of the item or component. Restricted physical or logical access. Physical access to the interior or exterior of the vehicle without the use of special tools. Examples include entering an unlocked vehicle and gaining access to an exposed interface such as the OBD port, or short-range wireless attacks that require the attacker to be close to the vehicle (e.g., replay attacks, attacks on key fob signals, and in-vehicle Wi-Fi).
- *Difficult:* Very low availability of the item or component. Impractical level of access to the item or component to perform the attack. Examples include decapping an integrated circuit and cracking a cryptographic key by brute force faster than the key can be rotated.

7.3.1.5 Equipment

Equipment refers to the technical equipment required by the attacker to discover the vulnerability and perform the attack. The rating according to ISO/SAE 21434, including examples, is described as follows:

- *Standard:* Equipment is readily available to the attacker. This equipment can be part of the product itself (e.g., a debugger in an operating system) or easily obtained (e.g., Internet sources, protocol analyzers, or simple attack scripts). Typical examples include a standard laptop, CAN dongle, and common tools and software (e.g., Kali Linux).

- *Specialized:* Equipment is not readily available to the attacker, but can be acquired without undue effort. This may involve the purchase of moderate amounts of equipment (e.g., power analysis tools, use of hundreds of PCs connected over the Internet would fall into this category), or the development of more sophisticated attack scripts or programs. If clearly different test benches consisting of specialized equipment are required for distinct steps of an attack, this would be rated as bespoke. Examples include specialized hardware debugging equipment, in-vehicle communications equipment, commercial tools, and specialized chemicals.
- *Bespoke:* Equipment is specially produced (e.g., very sophisticated software) and not readily available to the public (e.g., black market), or the equipment is so specialized that its distribution is controlled, perhaps even restricted. Or the equipment is very expensive. Examples include manufacturer-restricted tools and electron microscopes.
- *Multiple bespoke:* Situation where different types of bespoke equipment are required for the different steps of the attack.

7.3.2 CVSS-Based Approach

The Common Vulnerability Scoring System (CVSS) is a well-known open framework for rating the severity of software, and hardware vulnerabilities. We follow the CVSS version 3.1 [39].

CVSS consists of three metric groups called *base*, *temporal* and *environmental*. The *base score* is the most widely used score and represents the severity of the vulnerability according to its intrinsic characteristics. These characteristics are assumed to be constant over time, and the score is based on a worst-case impact for the various environments in which the vulnerability is present. The *temporal metrics* modify the base score representing characteristics that change over time, such as availability of the exploit code, and remediation level. Finally, the *environmental metrics* adjust both base and temporal scores to the characteristics of a specific environment. The combination of all applied groups results in the final severity score, which is a numerical score called *CVSS score*.

Limitation to Exploitability Metrics The CVSS scoring framework is designed to evaluate multiple aspects of a vulnerability that represent a risk. Within the base metrics, the characteristics of the vulnerabilities are rated by the *exploitability metrics*. In addition, the base metrics include the *score* and *impact metrics*. The score represents the impact of the vulnerability beyond its security scope. The impact metric, represents the impact of the vulnerability on the cybersecurity properties of confidentiality, integrity, and availability. ISO/SAE 21434 proposes to limit the evaluation of attack potential to the exploitability metrics only. The scope and impact metrics are not suitable in the context of determining the attack potential. For example, the impact of a vulnerability is modeled in ISO/SAE 21434 using a different approach based on damage scenarios (see *Annex G, Section G.3* in [2]). The temporal and environmental metrics are not suitable as well for the purposes of determining the attack potential of an attack path, and therefore they are not considered in this approach. It should be noted that by relying only on the exploitability metrics, the value calculated in this section is not the CVSS score according to the CVSS framework. Therefore, ISO/SAE 21434 calls this value the *CVSS exploitability value*.

The *exploitability metrics* are *attack vector (AV)*, *attack complexity (AC)*, *privileges required (PR)*, and *user interaction (UI)*. We describe each category shortly. A detailed description, and a calculator, can be found in [39].

7.3.2.1 Attack Vector (AV)

The possible values for rating the *attack vector (AV)* are:

- *Network (N)*: The target system (vulnerable component) can be remotely exploited from networks one or more hops away. This is the typical example of access via the Internet. Numerical value: 0.85.
- *Adjacent (A)*: The attacker must be on the same logical or physical network as the target system. Examples include a shared Wi-Fi or Bluetooth network, or a logical IP network, such as one accessed through a VPN. Numerical value: 0.62.
- *Local (L)*: The target system does not need to be connected to a network to exploit the vulnerability. The vulnerability is exploited either by gaining physical access to the system or by another person taking actions necessary to successfully exploit the vulnerability. In the latter case, the User Interaction value shall be considered. Numerical value: 0.55.
- *Physical (P)*: The attacker must make brief or persistent physical contact with the target system. Numerical value: 0.20.

7.3.2.2 Attack Complexity (AC)

The possible values for rating the *attack complexity (AC)* are:

- *Low (L)*: The attacker does not need any special condition on the target system to exploit the vulnerability. The attack will most likely succeed each time it is executed. Numerical value: 0.77.
- *High (H)*: The success of the exploit depends on conditions beyond the attacker's control. The attacker must prepare the target system before the vulnerability can be exploited. Examples include gathering information about the target system, including sensitive information such as configuration settings or shared secrets, or the attacker must gain access to a specific network. Numerical value: 0.44.

7.3.2.3 Privileges Required (PR)

The possible values for rating the *privileges required (PR)* are:

- *None (N)*: The attacker does not need any authorization on the target system to exploit the vulnerability. Numerical value: 0.85.
- *Low (L)*: Basic user privileges, such as settings and files that only affect the user, are required to exploit the vulnerability. Numerical value: 0.62 (Scope Unchanged).
- *High (H)*: The attacker must have higher privileges than a normal user. Numerical value: 0.27 (Scope Unchanged).

7.3.2.4 User Interaction (UI)

The possible values for rating the *user interaction (UI)* are:

- *None (N)*: The vulnerability can be exploited without user interaction. Numerical value: 0.85.
- *Required (R)*: User action is required before the vulnerability can be successfully exploited. Numerical value: 0.62.

7.3.2.5 CVSS Exploitability Value and Attack Feasibility Rating

The CVSS exploitability value ranges from 0.12 to 3.89, and it is calculated multiplying the numerical values of each rating category as follows:

Exploitability = 8.22 x AV x AC x PR x UI

The final attack feasibility rating for the attack path is determined by a mapping to the CVSS exploitability value. ISO/SAE 21434 provides an example of this mapping, which we show in Table 7.5.

Table 7.5 Mapping of the attack feasibility rating to the CVSS exploitability (as in *table G.8* in [2])

Attack feasibility rating	CVSS exploitability value
High	2.96–3.89
Medium	2.00–2.95
Low	1.06–1.99
Very low	0.12–1.05

Applicability of CVSS at TARA

The practical applicability of CVSS to automotive TARAs is, in our experience, very limited. The reason for this is that the attack paths typically consist of actions that cannot be easily modeled as vulnerabilities, as CVSS is designed to do. If we take the example of the attack path described in Sect. 7.2.2, many of the steps involved cannot be directly mapped to the categories of the CVSS exploitability metric. For example, if we want to rate the step *Rent a vehicle using falsified ID and credit card*, what rating should we apply to the Attack Vector? Since the step is a physical action in the real world, the only appropriate rating would be *physical*. Now, we have reached one of the biggest problems in applying CVSS to this type of step. While the value *physical* for the attack vector represents a very low exploitability (the numerical value for *physical* is 0.20), this means that regardless of the rest of the categories, the CVSS exploitability value will be below one, resulting in an attack feasibility rating of *very low*. A feasibility rating of very low for the step *renting a vehicle* is of course completely wrong because this is an action that anyone can do with very little effort, even if fake IDs and credit cards are used. To solve this problem, the evaluator must choose a higher value for the attack vector, such as *network*. The connection between *renting a vehicle* and *network* is not a natural approach to performing the evaluation. For this reason, the use of CVSS may not be practical. However, CVSS can be used to evaluate individual steps of an attack path that can actually be evaluated. In this case, it is entirely possible to mix the evaluation approaches, i.e., to use the attack-potential-based approach and the CVSS-based approach, depending on the characteristics of the step. However, additional matching of the resulting attack feasibility values is required, and the tool used to perform TARA must support this mixed approach.

7.3.3 *Attack Vector-Based Approach*

The attack vector-based approach calculates the attack feasibility rating based on only one parameter, the CVSS *attack vector (AV)*. This approach is recommended when the TARA is performed at the beginning of the development of the item or component and there is not enough technical information to apply other attack feasibility rating approaches. With the attack vector-based approach, it is at least possible to perform an initial TARA, identify the assets and relevant threats, and base the rating on the context in which the attack path can be exploited. When the product development reaches a certain maturity, the attack vector-based TARA can be improved and re-evaluated using, for example, the attack potential-based approach. An example of the mapping between the attack feasibility rating and the attack vector rating is shown in Table 7.6.

Table 7.6 Mapping of the attack feasibility rating to the Attack Vector of CVSS (as in *table G.9* in [2])

Attack feasibility rating	Attack Vector
High	Network
Medium	Adjacent
Low	Local
Very low	Physical

7.3.4 Determining the Attack Feasibility Rating by Attack Paths with Several Steps

An attack path usually consists of numerous steps, as we showed in our example in Sect. 7.2.2. The attack feasibility rating is calculated over the entire attack path, which means that in the end there is only one value, the attack feasibility rating, regardless of the number of steps required to complete the attack. While it is possible to determine the attack feasibility rating by reasoning about the feasibility of the attack path as a whole, especially when highly experienced cybersecurity teams are involved, it may be more accurate to evaluate the attack feasibility rating for each step individually. In the end, the attack feasibility rating for the entire attack path is equal to the combination of the maximum (less feasible) values of each category from each step involved.

Our example in Sect. 7.2.2 shows how the attack feasibility rating is determined when multiple steps are involved. We present the results in Table 7.7 using the attack potential-based approach. The maximum values for each category are highlighted. In this example, it can be observed that the first two steps, namely *Rent a vehicle using falsified ID and credit card* and *Open vehicle and extract head unit ECU* have the same total rating categories. The cybersecurity team can evaluate whether both steps are necessary or can be combined into one step.

The categories *knowledge of the item or component* and *window of opportunity* are the same for all steps in our example. For this reason, the final attack feasibility rating has these values for both categories, i.e., *public* and *unlimited* respectively. The category *elapsed time* has the largest value, i.e., less feasibility, by the step *Dump memory area where application code is stored and do reverse engineering on the code*. The value *≤6 months* is used for the final rating. *Specialist expertise* has also the highest rating by this step, namely the value *expert*. Note that we have highlighted the first highest value that appears for a category. The fact that other steps also have the value *expert* does not affect the final calculation. The category *equipment* has its maximum value, *specialized*, first in the step *Analyze which part of the code establishes communication to the back-end server and sends requests to a database*. The last column of Table 7.7 shows the attack feasibility rating calculated for each step. The combination of all values for each category, including the calculation of the final attack feasibility rating for the attack path, is summarized in the last row.

Table 7.7 Example of the calculation of the attack feasibility rating for an attack path with several steps (attack potential-based approach)

Steps of the attack path	Elapsed time	Specialist expertise	Knowledge	Window of oppor.	Equip-ment	Attack feasib.
Rent a vehicle using falsified ID and credit card	≤1 day	Layman	Public	Unlimited	Standard	High
Open vehicle and extract head unit ECU	≤1 day	Layman	Public	Unlimited	Standard	High
Research how to get diagnostics access to the ECU	≤1 week	Proficient	Public	Unlimited	Standard	High
Get necessary debugging tools and connect them to the ECU (e.g., using UDS, or unprotected JTAG)	≤1 week	Layman	Public	Unlimited	Standard	High
Dump memory area where application code is stored and do reverse engineering on the code	≤6 months	Expert	Public	Unlimited	Standard	Low
Analyze which part of the code establishes the communication to the back-end server and sends requests to a database	≤1 month	Expert	Public	Unlimited	Specialist	Medium
Manipulate the binary so a simple SQL injection string can be added to the database query	≤1 month	Expert	Public	Unlimited	Specialist	Medium
Overall rating for the attack path	**≤6 months**	**Expert**	**Public**	**Unlimited**	**Specialist**	**Very low**[1]

[1] The value *very low* is newly calculated from all categories of the row "Overall rating for the attack path"

It can be seen that the final attack feasibility rating is significantly lower (*very low*) than the individual steps involved.

As an example, we show the CVSS-based rating for the same example in Table 7.8. As discussed in Sect. 7.3.2, not all steps can actually be rated with CVSS.

Efficiency of Rating Multiple Steps As can be appreciated from the example in Table 7.7, rating the attack feasibility of each step, for each possible attack path,

Table 7.8 Example of the calculation of the attack feasibility rating for an attack path with several steps (CVSS-based approach)

Steps of the attack path	Attack Vector	Attack Complexity	Privileges Required	User Interaction	Attack feasib.
Rent a vehicle using falsified ID and credit card	–	–	–	–	–
Open vehicle and extract head unit ECU	–	–	-	–	–
Research how to get diagnostics access to the ECU	–	–	–	–	–
Get necessary debugging tools and connect them to the ECU (e.g., using UDS, or unprotected JTAG)	–	–	–	–	–
Dump memory area where application code is stored and do reverse engineering on the code	Physical	High	High	None	Very low
Analyze which part of the code establishes the communication to the back-end server and sends requests to a database	Physical	Low	None	None	Very low
Manipulate the binary so a simple SQL injection string can be added to the database query	Network	High	None	None	Medium
Overall rating for the attack path	**Network**	**High**	**High**	**None**	**Very low**

is a time-consuming activity. However, the evaluation is one of the most important parts of a TARA, and it must be performed rigorously to avoid having incorrect risk values at the end. In our experience, however, the efficiency of this evaluation improves with the experience of the cybersecurity team performing the TARA. The steps of an attack path depend on each item or component and cannot be generalized across projects. However, a highly experienced cybersecurity team may be able to perform the attack feasibility rating for an entire attack path. For example, the team can already estimate that the total score for the *elapsed time* must be *≤6 months* when reading the step *Dump memory area where application code is stored and do reverse engineering on the code*. Looking at the rest of the steps, the team concludes that there is no value higher than *≤6 months*. The same process applies to the rest of the categories.

7.3.5 *Relation to the Attacker Model*

We introduced the relationship between the attacker model and threats when we defined the attacker model as part of the scope in Sect. 5.6.3. We have defined a *threat level*, inspired by ETSI TVRA, which combines the capability with the motivation of an attacker. The attacker's capabilities represent the level of capability or attack potential that the attacker can achieve. Attacker capabilities are mapped to attack potential values according to the categories of the attack potential-based approach. For example, a threat level of *low* can represent an attacker with very high motivation but very low capability, or an attacker with very low motivation but formidable capability (cf. Table 5.4). The benefit of defining an attacker model and assigning attack potential levels to it is twofold. It provides the cybersecurity team with a deeper understanding of the threats and attack paths during the brainstorming session. This helps keep the focus on the threats and attack paths that are feasible for the particular TARA project. It also provides a systematic approach that can be directly compared to the attack feasibility evaluation. For example, if we define an attacker model with a threat level of *low* as described in Table 5.5, and in Table 5.6, we do not pursue attack paths on vehicles with a feasibility rating higher than *expert* for expertise, *public* for knowledge, and *standard* for equipment. A similar approach is followed in [36].

Note that the approach to threat levels and threat agents described in Table 5.5 and Table 5.6 is only one possible approach to attacker modeling. For example, we have not included elapsed time as a category for modeling the attacker. The list is also not exhaustive and should be considered as an example for didactic purposes, since we assume that the approach to modeling the attacker is specific to each organization. For example, we do not show all possible combinations of attack potential in Table 5.3. For example, it makes sense to discuss the question: *what happens if the attack path requires multiple experts, public knowledge, and standard equipment?* Table 5.3 can be expanded to cover several variations. A case requiring multiple experts, public knowledge, and standard equipment, could be mapped to the *limited* capability level.

7.4 Practical Case Studies

In Chap. 5, we introduced the three case studies and defined the scope of TARA. In Chap. 6, we identified assets, including damage scenarios, for each case study. In this section, we continue with the case studies. Based on the identified assets, we select a representative asset and conduct brainstorming sessions, supported by STRIDE, to identify threats. Due to space limitations, we have to limit ourselves to one asset and two threat scenarios per case study. However, we consider this sufficient for the purpose of explaining and understanding the process. We map the threats to the corresponding damage scenarios. Finally, we determine the attack path for one exemplary threat scenario, and we evaluate the attack feasibility rating.

7.4.1 Case Study 1: Automotive Embedded Device

We start the brainstorming session by focusing on one of the assets. We choose the asset *application software*. As described in Table 6.10, the application software is responsible for the video processing functions on one SoC and the connectivity functions on the other SoC. The identified cybersecurity properties for this asset are *integrity*, *availability*, and *authenticity*.

As explained in Sect. 7.1, we perform our threat modeling approach based on STRIDE and threat catalogs. In this example, we use the threat catalog from UN Regulation No. 155. Applying the threats from the regulation is a good approach to be compliant. The threats in UN Regulation No. 155 provide a basis for comparing and auditing TARAs. Since we do not rely on catalogs alone, we run STRIDE to infer other threats that are missing from the threat catalog.

To identify threats from the catalog that apply to our item, we follow these steps

1. Start with the first threat in the catalog. In the case of UN Regulation No. 155, the first threat is *1.1 Abuse of privileges by staff (insider attack)*, which belongs to the high-level threat *4.3.1 Threats regarding back-end servers related to vehicles in the field* and the sublevel threat *1 Back-end servers used as a means to attack a vehicle or extract data*. Note that we are not fussy about the wording. The catalog calls them *Example of vulnerability or attack method*, but they can be analyzed individually as separate threats. In fact, the exploited vulnerability or attack method is what we define as an attack path in our TARA, and attack paths must be described for each selected threat in the catalog anyway. The description in the regulation by the column *Example of vulnerability or attack method* is on the one hand too high level to apply as an attack path, and on the other hand the attack path is individual to the asset and item.
2. Is the threat technically feasible in the context of the asset and item? For example, the threat addresses cloud vulnerabilities, but we are analyzing embedded systems.
3. Is the threat within the scope of the TARA? For example, back-end servers are out of scope.
4. Does the threat compromise a cybersecurity property of the asset? Write down the reason why a cybersecurity asset is being compromised.
5. If the answers to the above questions are yes, the threat should be included in the TARA.
6. Finally, derive the attack path and perform a feasibility evaluation.

The following threats from the catalog of the UN Regulation No. 155 are examples that matche our asset *application software*:

- 5.1 Communications channels permit code injection, for example tampered software binary might be injected into the communication stream
- 5.2 Communications channels permit manipulate of vehicle held data/code
- 5.3 Communications channels permit overwrite of vehicle held data/code

- 5.4 Communications channels permit erasure of vehicle held data/code
- 5.5 Communications channels permit introduction of data/code to the vehicle (write data code)
- 6.1 Accepting information from an unreliable or untrusted source
- 12.1 Compromise of over the air software update procedures. This includes fabricating the system update program or firmware
- 12.2 Compromise of local/physical software update procedures. This includes fabricating the system update program or firmware
- 12.3 The software is manipulated before the update process (and is therefore corrupted), although the update process is intact
- 19.3 Extraction of cryptographic keys
- 22.2 Introduce malicious software or malicious software activity
- 23.1 Fabrication of software of the vehicle
- 26.2 Insufficient use of cryptographic algorithms to protect sensitive systems
- 26.3 Using already or soon to be deprecated cryptographic algorithms

For the sake of example, we analyze only one of the above threats in detail:

- *Threat:* 12.3 The software is manipulated before the update process (and is therefore corrupted), although the update process is intact
- *Attack path:*
  1. The attacker compromises the back-end server of the OEM / Tier (e.g., insider attacker)
  2. Manipulated software update is uploaded to the back-end server. The software does not have any cryptographic signature.
  3. The manipulated software is installed on the vehicle by normal update process (no cryptographic signature verification)

- *Compromised cybersecurity property:* Integrity, Availability
- *Reason for compromised cybersecurity property:*
  - Integrity: compromised due to software manipulation
  - Availability: manipulated update process and genuine update is not available

- *Attack feasibility rating:* Medium (see Table 7.9)

Assuming that the threats listed above are all possible threats identified from the catalog, we continue the analysis using STRIDE. The asset under analysis has three cybersecurity properties associated with it: *integrity*, *availability*, and *authenticity*. We can select the STRIDE categories that are directly related to these categories, namely spoofing (authenticity), tampering (integrity), and denial of service (availability). The approach at this point requires brainstorming and expertise from the cybersecurity team performing the TARA. For each of these three properties, the starting question for brainstorming would be *how can authenticity/integrity/availability be compromised?* Consider the following example for integrity and authenticity violations.

Table 7.9 Attack feasibility rating for threat scenario *12.3 The software is manipulated before the update process (and is therefore corrupted), although the update process is intact*

Steps of the attack path	Elapsed time	Specialist expertise	Knowledge	Window of oppor.	Equipment	Attack feasib.
The attacker compromises the back-end server of the OEM/Tier (e.g., insider attacker)	≤1 month	Proficient	Restricted	Unlimited	Standard	High
Manipulated software update is uploaded to the back-end server. The software does not have any cryptographic signature.	≤1 month	Expert	Restricted	Easy	Standard	Medium
The manipulated software is installed on the vehicle by normal update process (no cryptographic signature verification)	≤1 day	Layman	Public	Unlimited	Standard	High
Overall rating for the attack path	**≤1 month**	**Expert**	**Restricted**	**Easy**	**Standard**	**Medium**

- *Threat:* Bypass software signature verification by side channel attack
- *Attack path:*
  1. The attacker extracts the ECU from the vehicle and connects it in a local test bed with debugging capabilities
  2. The attacker launches side channel attacks based on power glitching
  3. The process for software signature verification is bypassed, and the attacker is able to install a modified software into the ECU

- *Compromised cybersecurity property:* Integrity, Authenticity
- *Reason for compromised cybersecurity property:*
  - Integrity: manipulated software is installed into the ECU
  - Authenticity: signature verification process bypassed

- *Attack feasibility rating:* Very low (see Table 7.10)

Note that the examples presented above, namely for the threat *12.3 The software is manipulated before the update process (and is therefore corrupted), although the update process is intact* and for the threat *Bypass software signature verification by side channel attack*, are based on different premises. In the first case, we assume that there are no controls in place, as it is actually defined in the scope of TARA (see Table 5.7). The ECU does not have a signature verification process and therefore cannot verify if the software has been tampered with. In the second example, we assume that there is a signature verification process in the ECU, but it is compromised. We have chosen this example to demonstrate STRIDE's methodology. Of course, at this

Table 7.10 Attack feasibility rating for threat scenario *Bypass software signature verification by side channel attack*

Steps of the attack path	Elapsed time	Specialist expertise	Knowledge	Window of oppor.	Equipment	Attack feasib.
The attacker extracts the ECU from the vehicle and connects it in a local test bed with debugging capabilities	≤1 day	Proficient	Public	Unlimited	Standard	High
The attacker launches side channel attacks based on power glitching	≤1 month	Expert	Public	Unlimited	Speciali.	Medium
The process for software signature verification is bypassed, and the attacker is able to install a modified software into the ECU	≤6 months	Expert	Public	Unlimited	Speciali.	Very low
Overall rating for the attack path	**≤6 months**	**Expert**	**Public**	**Unlimited**	**Speciali.**	**Very low**

point in the TARA project, the cybersecurity team may decide not to consider this threat because the initial assumption of TARA is that no controls are included. On the other hand, the threat should not be discarded because it will be helpful for a subsequent TARA for the same project.

7.4.2 Case Study 2: Vehicle-Backend Architecture

We use the same brainstorming methodology described earlier in this chapter to identify the threats in this case study. We begin the brainstorming session and select the first asset to analyze. We select the asset *sensitive data in transit* from the list of assets for this case presented in Table 6.14. The cybersecurity properties for this asset are *confidentiality*, *integrity*, and *authenticity*.

Moreover, we use the same threat catalog methodology combined with STRIDE that we used in Case Study 1 in Sect. 7.4.1. For this example, we start with the STRIDE analysis applied to the UN Regulation No. 155 catalog. Our main focus in the case studies is to give examples to better understand the methodology of finding threats in a TARA project. Therefore, we show how the STRIDE methodology can be applied and provide some examples. In a real project, if the TARA team decides to run STRIDE against a catalog, as shown in this case study, we would recommend that they not limit themselves to a specific catalog. For example, the team may want to run STRIDE on one or more catalogs to identify the most relevant threats more quickly, and then continue the threat identification process in a brainstorming session as a second step.

The asset *Sensitive data in transit* that we selected for analysis refers to data that has been labeled as sensitive by its data owner. This data can be, for example, personal names and credit card numbers being transmitted from the vehicle to a back-end system. The data can be stored in the vehicle and transferred to the back-end server using a direct connection to the back-end server over an Internet connection. In addition, the data can be transferred via Bluetooth to a trusted, locally paired cell phone that redirects the data to the back-end server using an application provided by the back-end owner, such as an OEM's iPhone® application. We explicitly chose to separate sensitive data into two assets to differentiate and show the importance of cybersecurity considerations at rest and in transit. In the context of this case study, in transit means end-to-end from the vehicle to the back-end system. It is perfectly valid to model both assets as a single asset and derive cybersecurity goals and cybersecurity controls for the entire context (locally stored and in transit).

The cybersecurity properties for the asset *sensitive data in transit* are *confidentiality*, *integrity*, and *authenticity*. As we described in Table 7.1, the STRIDE threat categories that are the opposite of these cybersecurity properties are *information disclosure*, *tampering*, and *spoofing*, respectively. One approach to applying STRIDE to the UN Regulation No. 155 catalog would be to go through the catalog's list of threats and ask *can this threat be generically related to information disclosure, tampering, or spoofing?*. If the answer is yes, the threat can be analyzed in more detail and applied to the TARA if it is a reasonable threat. The following threats are examples from the UN Regulation No. 155 catalog that we have identified for the *Sensitive data in transit* asset using the STRIDE approach:

- 3.1 Abuse of privileges by staff (insider attack)
- 3.2 Loss of information in the cloud. Sensitive data may be lost due to attacks or accidents when data is stored by third-party cloud service providers
- 3.3 Unauthorized Internet access to the server (enabled for example by backdoors, unpatched system software vulnerabilities, SQL attacks or other means)
- 3.4 Unauthorized physical access to the server (conducted for example by USB sticks or other media connecting to the server)
- 5.1 Communications channels permit code injection, for example tampered software binary might be injected into the communication stream
- 5.2 to 5.4 Communications channels permit manipulate/overwrite/erasure/write of vehicle held data/code
- 6.2 Man in the middle attack/session hijacking
- 6.3 Replay attack, for example an attack against a communication gateway allows the attacker to downgrade software of an ECU or firmware of the gateway
- 7.2 Gaining unauthorized access to files or data

- 9.1 An unprivileged user is able to gain privileged access, for example root access
- 17.1 Corrupted applications, or those with poor software security, used as a method to attack vehicle systems
- 18.1 External interfaces such as USB or other ports used as a point of attack, for example through code injection
- 19.3 Extraction of cryptographic keys
- 23.1 Fabrication of software of the vehicle
- 26.1 Combination of short encryption keys and long period of validity enables attacker to break encryption
- 29.1 Superfluous Internet ports left open, providing access to network systems
- 31.1 Information breach. Personal data may be leaked when the car changes user (e.g., is sold or is used as hire vehicle with new hirers)
- 32.1 Manipulation of electronic hardware, e.g., unauthorized electronic hardware added to a vehicle to enable 'man-in-the-middle' attack; Replacement of authorized electronic hardware (e.g., sensors) with unauthorized electronic hardware; Manipulation of the information collected by a sensor (for example, using a magnet to tamper with the Hall effect sensor connected to the gearbox)

After the threats are identified in the brainstorming session (i.e., after several rounds of discussion using STRIDE and catalogs), the threat scenarios are described in detail, specifically explaining the compromised cybersecurity properties of the asset and the attack path involved. One of the identified threats is described below:

- *Threat:* 6.2 Man in the middle attack/session hijacking
- *Attack path:*

  1. Assuming that the Bluetooth connection between the cell phone and the vehicle does have vulnerabilities (i.e., not enough security features)
  2. The attacker knows about the Bluetooth configuration used by a certain vehicle platform and the attacker buys or rents one car, in order to perform enough tests, and research
  3. The attacker builds an own device, which is easy to transport and can be used to perform a MITM attack on Bluetooth
  4. The attacker gets close enough to vulnerable vehicles and performs the attack

- *Compromised cybersecurity property:* Confidentiality
- *Reason for compromised cybersecurity property:*

  – Confidentiality: the attacker performs eavesdropping of sensitive data

- *Attack feasibility rating:* Medium (see Table 7.11)

Table 7.11 Attack feasibility rating for threat scenario *6.2 Man in the middle attack/session hijacking*

Steps of the attack path	Elapsed time	Specialist expertise	Knowledge	Window of oppor.	Equipment	Attack feasib.
The attacker knows about the Bluetooth configuration used by a certain vehicle platform and the attacker buys or rents one car, in order to perform enough tests, and research	≤1 month	Proficient	Public	Unlimited	Specialist	High
The attacker builds an own device, which is easy to transport and can be used to perform a MITM attack on Bluetooth	≤1 week	Proficient	Public	Unlimited	Specialist	High
The attacker gets close enough to vulnerable vehicles and performs the attack	≤1 day	Proficient	Public	Moderate	Specialist	High
Overall rating for the attack path	**≤1 month**	**Proficient**	**Public**	**Moderate**	**Specialist**	**Medium**

7.4.3 *Case Study 3: Automotive Manufacturing Plant*

To identify the threats to the automotive manufacturing plant, we use the same approach of brainstorming sessions supported by STRIDE and catalogs that we used in the previous case studies. The identification begins by analyzing each of the assets already identified in Table 6.19. We choose the asset *software updates* as an example for our threat identification process. The assigned cybersecurity properties for this asset are *integrity* and *authenticity*.

The asset *software updates* in our case study represent updates for the production machines. A typical manufacturing plant in the automotive sector may include robots, conveyors, PCB assembly machines, CNC machines, soldering machines, material mixing machines, compressors, pumps, cleaners, packaging machines, functional tests, thermomechanical tests, EOL tests, and others. In our case study, we analyze a manufacturing facility consisting of a production line with several stations, such as a flash station, some machines in between, and an EOL station. We assume that each machine involved in the production line has some degree of software update capability, so the asset *software updates* include all of them. At this point, it is important to remember that this case study aims to identify risks at an architectural or system level. Certainly, we are taking some shortcuts in our example. At the system level, the brainstorming team focuses on the possible threats and attack vectors that are feasible for the machines as a whole. For example, if none of the machines have remote update capabilities, the threat vectors involving Ethernet attacks will not apply. In this case, the threats will be related to attacks that involve physical access to the machines (e.g., introducing USB sticks, performing social engineering attacks, accessing configuration interfaces). Of course, if the TARA project requires detailed insight into the individual machines, the asset *software updates* can be specified for each individual machine.

The brainstorming session begins with the threat catalogs that STRIDE supports. The cybersecurity team begins by analyzing the cybersecurity property *integrity*. There are enough experts in the brainstorming session who know the details of the machine, at least from an operational perspective. The experts can confirm, for example, if the machines can be updated using remote services over Ethernet, such as *HTTP*, *SMB*, *FTP*, *SSH*, or *telnet*, or if the updates are only possible by physical access. The process can begin by following the UN Regulation No. 155 catalog and reasoning whether the threats in the catalog may apply to the asset. Following the same approach as in Case Study 2, STRIDE helps to understand if the threats are of the type described as tampering threats that compromise the integrity of the asset. Since we are analyzing an OT environment, we also evaluate the industrial control system threats that exist in the MITRE ATT&CK® framework. At the end of the catalog process, the cybersecurity team continues to run STRIDE, identifying every possible manipulation threat from the expertise and imagination of the experts involved. The process is repeated for the cybersecurity property *authenticity*.

Since the machines in the production plant support local and remote updates, we have identified the following threats from UN Regulation No. 155:

- 3.1 Abuse of privileges by staff (insider attack)
- 3.3 Unauthorized Internet access to the server (enabled for example by backdoors, unpatched system software vulnerabilities, SQL attacks or other means)
- 3.4 Unauthorized physical access to the server (conducted for example by USB sticks or other media connecting to the server)
- 5.1 Communications channels permit code injection, for example tampered software binary might be injected into the communication stream
- 12.2 Compromise of local/physical software update procedures. This includes fabricating the system update program or firmware
- 22.2 Introduce malicious software or malicious software activity
- 29.1 Superfluous Internet ports left open, providing access to network systems
- 29.2 Circumvent network separation to gain control

Note that the attack paths for the UN Regulation No. 155 threats applied to an OT device are significantly different from the attack paths when the device is an ECU, as in our first case study.

The following threats are examples of threats from the ICS catalog of the MITRE ATT&CK® (extracted from [13]):

- Command-Line Interface (T0807): Adversaries may utilize command-line interfaces (CLIs) to interact with systems and execute commands. CLIs provide a means of interacting with computer systems and are a common feature across many types of platforms and devices within control systems environments. Adversaries may also use CLIs to install and run new software, including malicious tools that may be installed over the course of an operation. CLIs are typically accessed locally, but can also be exposed via services, such as SSH, telnet, and RDP. Commands that are executed in the CLI execute with the current permissions level of the process running the terminal emulator, unless the command specifies a change in permissions context. Many controllers have CLI interfaces for management purposes.
- Default Credentials (T0812): Adversaries may leverage manufacturer or supplier set default credentials on control system devices. These default credentials may have administrative permissions and may be necessary for initial configuration of the device. It is general best practice to change the passwords for these accounts as soon as possible, but some manufacturers may have devices that have passwords or usernames that cannot be changed.
- Exploitation of Remote Services (T0866): Adversaries may exploit a software vulnerability to take advantage of a programming error in a program, service, or within the operating system software or kernel itself to enable remote service abuse. A common goal for post-compromise exploitation of remote services is for initial access into and lateral movement throughout the ICS environment to enable access to targeted systems.
- System Firmware (T0857): System firmware on modern assets is often designed with an update feature. Older device firmware may be factory installed and require special reprogramming equipment. When available, the firmware update feature enables vendors to remotely patch bugs and perform upgrades. Device firmware updates are often delegated to the user and may be done using a software update package. It may also be possible to perform this task over the network. An adversary may exploit the firmware update feature on accessible devices to upload malicious or out-of-date firmware. Malicious modification of device firmware may provide an adversary with root access to a device, given firmware is one of the lowest programming abstraction layers.
- Remote Services (T0886): Adversaries may leverage remote services to move between assets and network segments. These services are often used to allow operators to interact with systems remotely within the network, some examples are RDP, SMB, SSH, and other similar mechanisms.
- Hardcoded Credentials (T0891): Adversaries may leverage credentials that are hardcoded in software or firmware to gain an unauthorized interactive user session to an asset. Examples of credentials that may be hardcoded in an asset include:

username/passwords, cryptographic keys/certificates, and API tokens. These credentials are built into the system in a way that they either cannot be changed by the asset owner, or may be infeasible to change because of the impact it would cause to the control system operation. These credentials may be reused across whole product lines or device models, and are often not published or known to the owner and operators of the asset. Adversaries may utilize these hardcoded credentials to move throughout the control system environment or provide reliable access for their tools to interact with industrial assets.

Finally, we perform the description of all threat scenarios for the selected asset, including the attack path. We describe one example as follows:

- *Threat:* System Firmware (T0857)
- *Attack path:*

  1. The attacker is an insider who knows the vendor and model version of the safety PLCs used in the victim's manufacturing facility
  2. The attacker prepares USB sticks containing the malware *Triton* [40]. The USB stick is carefully labeled so that an innocent victim could believe it is a regular USB containing information for each standard step in the production process.
  3. The attacker distributes the USB sticks within the department that normally maintains the systems
  4. An innocent operator inserts the USB stick into the same network as the PLCs and the malware is executed (e.g., by hiding it in some auto-start script)
  5. The malware installs a modified firmware into the target PLCs

- *Compromised cybersecurity property:* Integrity, Authenticity
- *Reason for compromised cybersecurity property:*

  - Integrity: compromised because of a manipulation of software
  - Authenticity: the malware in the USB stick was executed without any authenticity check. Additionally, the malware was able to access to the PLCs.

- *Attack feasibility rating:* Medium (see Table 7.12)

Table 7.12 Attack feasibility rating for threat scenario *System Firmware (T0857)*

Steps of the attack path	Elapsed time	Specialist expertise	Knowledge	Window of oppor.	Equipment	Attack feasib.
The attacker is an insider who knows the vendor and model version of the safety PLCs used in the victim's manufacturing facility	≤1 day	Expert	Confid.	Unlimited	Standard	High
The attacker prepares USB sticks containing the malware *Triton* [40]. The USB stick is carefully labeled so that an innocent victim could believe it is a regular USB containing information for any standard step of the production process.	≤1 month	Expert	Public	Unlimited	Standard	High
The attacker distributes the USB sticks within the department that normally maintains the systems	≤1 day	Layman	Restricted	Moderate	Standard	High
An innocent operator inserts the USB stick into the same network as the PLCs and the malware is executed (e.g., by hiding it in some auto-start script)	≤1 day	Proficient	Restricted	Moderate	Standard	High
The malware installs a modified firmware into the target PLCs	≤1 day	Layman	Public	Moderate	Standard	High
Overall rating for the attack path	**≤1 month**	**Expert**	**Restricted**	**Moderate**	**Standard**	**Medium**

References

1. OWASP Top Ten—OWASP Foundation. https://owasp.org/www-project-top-ten/
2. ISO/SAE 21434:2021 Road vehicles—Cybersecurity engineering. Standard, International Organization for Standardization/Society of Automotive Engineers, Geneva, CH (2021). https://www.iso.org/standard/70918.html
3. Boudguiga, A., Boulanger, A., Chiron, P., Klaudel, W., Labiod, H., Seguy, J.C.: RACE: Risk analysis for cooperative engines. In: 2015 7th International Conference on New Technologies, Mobility and Security (NTMS), pp. 1–5 (2015). https://doi.org/10.1109/NTMS.2015.7266516
4. ThreatModeler: Threat Modeling Methodologies: What is VAST? https://threatmodeler.com/threat-modeling-methodologies-vast/

5. Macher, G., Sporer, H., Berlach, R., Armengaud, E., Kreiner, C.: SAHARA: A security-aware hazard and risk analysis method. In: 2015 Design, Automation & Test in Europe Conference & Exhibition (DATE), pp. 621–624 (2015). https://doi.org/10.7873/DATE.2015.0622
6. Alberts, C., Behrens, S., Pethia, R., Wilson, W.: Operationally Critical Threat, Asset, and Vulnerability Evaluation (OCTAVE) Framework, Version 1.0. Tech. Rep. CMU/SEI-99-TR-017, Software Engineering Institute, Carnegie Mellon University, Pittsburgh, PA (1999). http://resources.sei.cmu.edu/library/asset-view.cfm?AssetID=13473
7. Deng, M., Wuyts, K., Scandariato, R., Preneel, B., Joosen, W.: A Privacy Threat Analysis Framework: Supporting the Elicitation and Fulfillment of Privacy Requirements. Requir. Eng. 16(1), 3–32 (2011). https://doi.org/10.1007/s00766-010-0115-7
8. Luo, F., Jiang, Y., Zhang, Z., Ren, Y., Hou, S.: Threat analysis and risk assessment for connected vehicles: A survey. Security and Communication Networks 2021, 1–19 (2021). https://doi.org/10.1155/2021/1263820. https://www.hindawi.com/journals/scn/2021/1263820/
9. Shostack, A.: Threat Modeling: Designing for Security, 1st edn. Wiley Publishing (2014)
10. Threat Modeling Manifesto. http://www.threatmodelingmanifesto.org/
11. Upstream: Upstream 2022 Global Automotive Cybersecurity Report (2022). https://upstream.auto/2022report/
12. United Nations: UN Regulation No. 155—Cyber security and cyber security management system E/ECE/TRANS/505/Rev.3/Add.154 (2021). https://unece.org/transport/documents/2021/03/standards/un-regulation-no-155-cyber-security-and-cyber-security
13. The MITRE Corporation: MITRE ATT&CK®. https://attack.mitre.org/
14. European Union Agency for Cybersecurity (ENISA): ENISA Threat Landscape 2022 (2022). https://www.enisa.europa.eu/publications/enisa-threat-landscape-2022
15. European Union Agency for Cybersecurity (ENISA): ENISA good practices for security of Smart Cars (2019). https://www.enisa.europa.eu/publications/smart-cars
16. Automotive Information Sharing and Analysis Center (Auto-ISAC) (2023). https://automotiveisac.com
17. Auto-ISAC Automotive Threat Matrix (ATM) (2024). https://atm.automotiveisac.com/
18. Automotive Security Research Group (ASRG) (2023). https://asrg.io
19. ISO/IEC 27005:2022 Information security, cybersecurity and privacy protection—Guidance on managing information security risks. Standard, International Organization for Standardization/International Electrotechnical Commission (2022). https://www.iso.org/standard/80585.html
20. Kohnfelder, L., Garg, P.: The threats to our products. Microsoft Interface, Microsoft Corporation (1999)
21. Muckin, M., Fitch, S.C.: A Threat-Driven Approach to Cyber Security . White paper, Lockheed Martin Corporation (2019). https://www.lockheedmartin.com/content/dam/lockheed-martin/rms/documents/cyber/LM-White-Paper-Threat-Driven-Approach.pdf
22. EVITA Project: E-safety vehicle intrusion protected applications (EVITA). https://www.evita-project.org/
23. SAE J3061—Cybersecurity Guidebook for Cyber-Physical Vehicle Systems. Standard, Society of Automotive Engineers (2021). https://www.sae.org/standards/content/j3061_202112/?src=j3061_201601
24. Ruddle, A., Ward, D., Weyl, B., Idrees, S., Roudier, Y., Friedewald, M., Leimbach, T., Fuchs, A., Gürgens, S., Henniger, O., Rieke, R., Ritscher, M., Broberg, H., Apvrille, L., Pacalet, R., Pedroza, G.: Security requirements for automotive on-board networks based on dark-side scenarios (EVITA Deliverable 2.3) (2009). https://doi.org/10.5281/zenodo.1188418
25. ISO/IEC 18045:2008 Information technology—Security techniques—Methodology for IT security evaluation. Standard, International Organization for Standardization/International Electrotechnical Commission (2008). https://www.iso.org/standard/46412.html
26. ISO 26262:2018 Road vehicles - Functional safety, Part 1-12. Standard, International Organization for Standardization (2018)
27. Kelling, E., Friedewald, M., Leimbach, T., Menzel, M., Säger, P., Seudié, H., Weyl, B.: Specification and evaluation of e-security relevant use cases (EVITA Deliverable 2.1) (2018). https://doi.org/10.5281/zenodo.1188415

28. Lautenbach, A., Mafijul Islam, C.: HEAling vulnerabilities to ENhance software security and safety (HEAVENS)—deliverable d2—security models (2016). https://autosec.se/wp-content/uploads/2018/03/HEAVENS_D2_v2.0.pdf
29. Lautenbach, A., Almgren, M., Olovsson, T.: Proposing HEAVENS 2.0—an Automotive Risk Assessment Model. In: Proceedings of the 5th ACM Computer Science in Cars Symposium, CSCS '21. Association for Computing Machinery, New York, NY, USA (2021). https://doi.org/10.1145/3488904.3493378
30. GitLab: Threat Modeling. https://handbook.gitlab.com/handbook/security/product-security/application-security/threat-modeling/
31. VerSprite: PASTA Threat Modeling: The Process for Attack Simulation and Threat Analysis (eBook) (2022). https://versprite.com/cybersecurity-listings/offsec/pasta-threat-modeling/
32. Tony UcedaVelez: Real World Threat Modeling Using the PASTA Methodology, OWASP AppSec EU 2012 (2012). https://wiki.owasp.org/images/a/aa/AppSecEU2012_PASTA.pdf
33. Schneier, B.: Attack Trees, Dr. Dobb's Journal, December 1999 (1999). https://www.schneier.com/academic/archives/1999/12/attack_trees.html
34. Common Methodology for Information Technology Security Evaluation—Evaluation methodology. CEM:2022, Revision 1, CCMB-2022-11-006. Common Criteria. Standard, CCRA Members (2022). https://www.commoncriteriaportal.org/files/ccfiles/CEM2022R1.pdf
35. ETSI: ETSI TS 102 165-1 V5.2.5—CYBER; Methods and protocols; Part 1: Method and pro forma for Threat, Vulnerability, Risk Analysis (TVRA). Standard (2022). https://www.etsi.org/deliver/etsi_ts/102100_102199/10216501/05.02.05_60/ts_10216501v050205p.pdf
36. Zelle, D., Plappert, C., Rieke, R., Scheuermann, D., Krauß, C.: ThreatSurf: A method for automated Threat Surface assessment in automotive cybersecurity engineering. Microprocessors and Microsystems 90, 104461 (2022). https://doi.org/10.1016/j.micpro.2022.104461. https://www.sciencedirect.com/science/article/pii/S0141933122000321
37. ISO/IEC 18045:2022 Information security, cybersecurity and privacy protection—Evaluation criteria for IT security—Methodology for IT security evaluation. Standard, International Organization for Standardization / International Electrotechnical Commission (2022). https://www.iso.org/standard/72889.html
38. Smith, C.: The Car Hacker's Handbook: A Guide for the Penetration Tester, 1 edn. No Starch Press, San Francisco (2016)
39. Common Vulnerability Scoring System (CVSS) v3.1 Specification Document—Revision 1. Specification document, Forum of Incident Response and Security Teams (FIRST) (2019). https://www.first.org/cvss/v3-1/cvss-v31-specification_r1.pdf
40. The MITRE Corporation: Triton, Software S1009—MITRE ATT&CK®. https://attack.mitre.org/versions/v12/software/S1009/

Chapter 8
Performing the Risk Evaluation

Abstract This chapter addresses the calculation of risk. Risk is derived from the combination of the feasibility rating of the threat scenarios and the impact of the threat scenario actually occurring. We perform the mapping of asset-threat scenarios to damage scenarios and calculate the risk of each combination. We provide practical insights on how to adjust the risk matrix. We also show how statistical representations can be useful to better present and understand the key results of a TARA. Finally, we provide examples based on our three case studies.

In this chapter, we have come to the essential task of threat analysis and risk assessment, which is the calculation of risk. Risk is derived by combining the feasibility rating of the threat scenarios with the impact of the threat scenario actually occurring. ISO/SAE 21434 assigns a numerical value to risk, ranging from one (minimum risk) to five, and can be determined using a risk matrix or risk formula. The risk evaluation step in the overall TARA process is depicted in Fig. 8.1.

Necessary input for this chapter:

- Threat scenarios
- Feasibility evaluation
- Damage scenarios with impact ratings

The outcome of this chapter is:

- Risk value for each combination of threat scenario

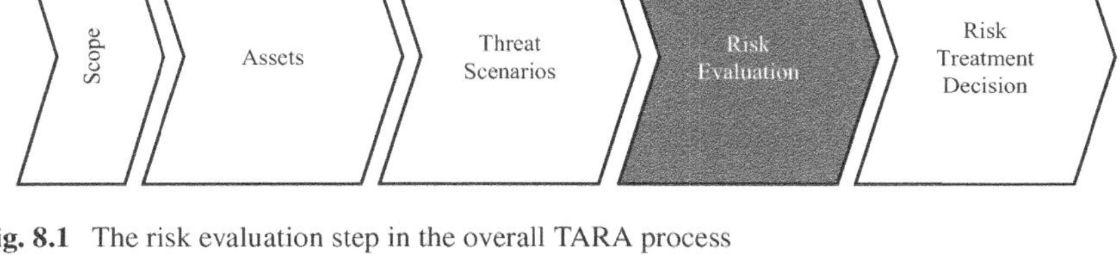

Fig. 8.1 The risk evaluation step in the overall TARA process

R. do Carmo and A. Schlensog, *Automotive Threat Analysis and Risk Assessment in Practice*, https://doi.org/10.1007/978-3-662-69614-9_8

8.1 Risk Evaluation

Once the threat scenarios have been assessed for feasibility and the impact of the damage scenarios has been quantified, determining the risk value is straightforward. In this step, the threat scenarios and the resulting damage scenarios are combined. Regardless of the tool used to perform the TARA, the combination of assets, threat scenarios, including different attack paths, and the resulting damage scenarios must be created. When everything is put into a matrix, the resulting TARA looks like the matrix shown in Fig. 8.2. Each asset in the item has one or more threat scenarios associated with it. As explained in Chap. 7, the relationship between asset and threat is primarily modeled through cybersecurity properties. Then, each threat scenario has at least one attack path that realizes the threat scenario in the context of the asset. Note that in Fig. 8.2 we explicitly show the different attack paths for the same threat scenario *threat scenario 1* to better understand how the process is thought out. The ISO/SAE 21434 allows combining all attack paths for the same threat scenario by performing an aggregation of the feasibility ratings. In other words, the maximum of the attack feasibility ratings of all attack paths is assigned to the threat scenario. We show the combined attack path for *threat scenario 2*. Finally, the risk score is determined for each combination of asset-threat-scenario-damage scenario. ISO/SAE 21434 specifies that the risk value should be a numeric value ranging from one to five. However, organizations have the flexibility to create their own risk matrices or formulas to calculate and assign the risk value based on their specific requirements and preferences.

Do not confuse attack paths with steps of attack paths
If a threat scenario is realized by multiple attack paths, the risk calculation is performed by analyzing only the simplest attack path, i.e., the attack path with the highest attack feasibility rating. This is a reasonable approach because, eventually, the risk resulting from the threat scenario must be treated regardless of the attack paths. The simplest attack path carries the highest risk. The procedure for selecting the simplest attack path is similar to the procedure for determining the feasibility of attack paths consisting of multiple steps (see Sect. 7.3.4). However, they are different procedures and should not be confused. We illustrate the difference and the relationship between multiple steps of attack paths and multiple attack paths for a threat scenario in Fig. 8.3.

8.1.1 Risk Matrix

The risk value can be calculated using a risk matrix such as the one shown in Table 8.1. For example, if a certain damage scenario has an impact rating of *moderate*, and

Asset information		Threat scenario information					Damage scenario infomation			Risk evaluation	
Asset [WP-15-02]	Cybersecurity property	Threat scenario [WP-15-03]	Attack path [WP-15-05]	Compromised cybersecurity property of the asset [RQ-15-03]	Cause of compromise of the cybersecurity property [RQ-15-03]	Attack feasibility rating [WP-15-06]	Damage scenario [WP-15-01]	Impact rating for road user [WP-15-04]	Impact rating for OEM / Supplier	Risk value for road user [WP-15-07]	Risk value for OEM / Supplier
Relationship Assets → Attack path → Threat scenarios											
Asset 1	Confidentiality, Integrity, Availability	Threat scenario 1	Attack path 1	Integrity, Availability	…	…	Damage scenario 1	Safety: … Financial: … Operational: … Privacy: …	…	Safety: … Financial: … Operational: … Privacy: …	…
							Damage scenario 2	Safety: … Financial: … Operational: … Privacy: …	…	Safety: … Financial: … Operational: … Privacy: …	…
			Attack path 2	Integrity, Availability	…	…	Damage scenario 1	Safety: … Financial: … Operational: … Privacy: …	…	Safety: … Financial: … Operational: … Privacy: …	…
							Damage scenario 2	Safety: … Financial: … Operational: … Privacy: …	…	Safety: … Financial: … Operational: … Privacy: …	…
		Threat scenario 2	Attack path (easiest path) (Maximum feasibility rating over all possible attack paths)	Confidentiality	…	…	Damage scenario 1	Safety: … Financial: … Operational: … Privacy: …	…	Safety: … Financial: … Operational: … Privacy: …	…
							Damage scenario 2	Safety: … Financial: … Operational: … Privacy: …	…	Safety: … Financial: … Operational: … Privacy: …	…
							Damage scenario 3	Safety: … Financial: … Operational: … Privacy: …	…	Safety: … Financial: … Operational: … Privacy: …	…

Fig. 8.2 Depiction of the risk evaluation as a matrix including its relationship to assets, threat scenarios, and attack paths

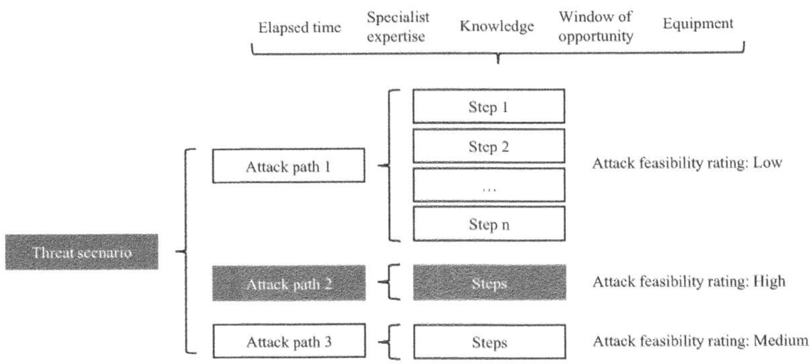

Fig. 8.3 Difference between multiple attack paths and multiple steps of attack paths. The three attack paths consist of multiple steps. Attack path two has the highest attack feasibility rating and is selected for risk evaluation

Table 8.1 Example of risk matrix (as in *table H.8* in [1])

Impact rating	Attack feasibility rating			
	Very low	**Low**	**Medium**	**High**
Severe	2	3	4	5
Major	1	2	3	4
Moderate	1	2	2	3
Negligible	1	1	1	1

the corresponding threat scenario has an attack feasibility rating of *high*, then the resulting risk value is three.

8.1.1.1 Adaptations to the Risk Matrix

The organization can make adjustments to the risk matrix (or formula) used. If we analyze the example of the risk matrix in Table 8.1, we can easily see that the risks resulting from a *negligible* impact always have the minimum risk value one. This mapping is reasonable if the impact rating categories for *negligible* always represent an irrelevant consequence, such as those already presented in Table 6.3. If the organization uses its own impact rating criteria, and these criteria have a higher impact for *negligible* than a completely irrelevant consequence, it may be reasonable to modify the risk matrix to assign a higher risk to threats that have a *negligible* impact but occur too often. For example, if the *negligible* criterion for financial impact on the road user represents a $1 loss to the road user every time the threat is successfully executed, the risk to the road user may be very low if the feasibility of the threat is also very low. In effect, it means that it is very unlikely that the threat will ever be executed, and even if it is, the road user will lose $1. On the other hand, if the threat has a high attack feasibility rating, i.e., it is very likely that the threat will be executed, the risk to the road user is much higher. For example, if the threat is executed 10,000 times, the road user will lose $10,000.

A similar challenge occurs when the threat scenario is associated with a very high-impact rating (severe), but its likelihood of occurrence is very low. The problem in this case is that it is not possible to reduce the risk to a minimal level. Even if we take an extreme example for the attack feasibility rating, such as when a group of several experts must brute-force AES-256, the risk level will still be two when using the risk matrix in Table 8.1. We show an example of an adjusted risk matrix in Table 8.2.

8.1.2 Risk Formula

The calculation of the risk value can be performed using a risk formula. For example, the ISO/SAE 21434 uses the following example,

$$R = 1 + I x F$$

where R is the risk value, I a numeric value for the impact rating, and F a numeric value for the attack feasibility rating. The translation of the impact rating and attack feasibility rating into numeric values are depicted in Table 8.3 and in Table 8.4, respectively.

Table 8.2 Example of a risk matrix. Negligible impact becomes a higher risk value when the risk likelihood is higher. Attacks with a very low likelihood represent no considerable risk.

Impact rating	**Attack feasibility rating**			
	Very low	**Low**	**Medium**	**High**
Severe	1	3	4	5
Major	1	3	4	5
Moderate	1	2	3	4
Negligible	1	1	2	3

Table 8.3 Example translation of impact rating to numerical values (as in *table H.10* in [1])

Impact rating	I
Negligible	0
Moderate	1
Major	1.5
Severe	2

Table 8.4 Example translation of attack feasibility rating to numerical values (as in *table H.10* in [1])

Attack feasibility rating	***F***
Very low	0
Low	1
Medium	1.5
High	2

8.2 Understanding Data and Working with Statistics

Each individual risk identified is important and needs to be addressed. However, it is often useful to work with statistics to understand the big picture. The individual results of the risk evaluation can be useful when going deeper into the engineering process and, for example, tracing risks back to controls. In real-world projects, the cybersecurity team will most likely need to present the results of the TARA to project management or upper management. The cybersecurity team may need to present the most relevant results of the TARA, such as the most relevant risks, most affected assets, or most occurring threat scenarios. The main reason for working with statistics is that a TARA usually consists of many risks. The risk evaluation matrix shown in Fig. 8.2 is only a minimal example of how the risks in a TARA are represented by the combination of their respective assets and threat scenarios. Each row of the TARA with an associated risk corresponds to a combination of asset-threat scenario-damage scenarios. The number of rows, i.e., the number of risk values, is ultimately equal to the multiplication of

Number of assets *x* Number of threat scenarios *x* Number of damage scenarios

(the formula is simplified, the threat scenarios and damage scenarios vary for each asset). For example, a real-world TARA containing an item with 20 assets, an average of ten threat scenarios per asset, and an average of five damage scenarios per threat scenario, will have 1,000 rows.

8.2.1 Threat Scenarios

An example of a statistical analysis for threat scenarios from a TARA is shown in Fig. 8.4. The analysis shows the number of asset-threat-scenario-damage-scenario combinations for all threat scenarios used for the TARA. The occurrences are shown per risk level, making it easy to identify critical threats. The total number of occurrences is, of course, an interesting parameter for understanding which threats are most prevalent. The total number of occurrences does not necessarily correspond to higher risk levels. For example, the first threat scenario in Fig. 8.4 has the largest number of occurrences (469), but it has no risk at severity level five. The risk levels

Threat scenario	Statistics for threat scenarios — Count occurrences *Asset-Threat scenario-Damage scenario*						
	Total	Occurrences per risk level					
		5	4	3	2	1	Risks > 2
16.2 Manipulation of vehicle telematics	469	0	60	189	220	0	249
12.3 The software is manipulated before the [...]	453	200	178	63	12	0	441
12.2 Compromise of local/physical software [...]	448	210	160	70	8	0	440
5.1 Communications channels permit code inj. [...]	380	187	124	66	0	3	377
23.1 Fabrication of software of the vehicle	362	223	104	20	7	8	347
29.1 Superfluous Internet ports left open, [...]	278	122	98	30	23	5	250
18.3 Diagnostic access	259	102	98	54	5	0	254
18.2 Media infected with a virus connected to [...]	245	10	35	168	30	2	213
8 Denial of service attacks via communication [...]	195	5	63	78	36	13	146
18.1 External interfaces such as USB or other [...]	185	98	45	30	12	0	173
4.1 Spoofing of messages by impersonation	167	104	63	0	0	0	167
11.3 Malicious diagnostic messages	163	89	74	0	0	0	163
11.4 Malicious proprietary messages	148	63	53	32	0	0	148
6.1 Accepting information from an unreliable [...]	127	89	36	0	0	2	125
16.3 Interference with short range wireless [...]	114	10	87	0	12	5	97
16.1 Manipulation of functions designed to [...]	112	89	23	0	0	0	112
20.5 Unauthorized changes to system diag. [...]	109	56	43	0	10	0	99
13.1 Denial of Service attack against update [...]	47	0	47	0	0	0	47
19.3 Extraction of cryptographic keys	42	30	8	4	0	0	42
17.1 Corrupted applications, or those with poor [...]	36	24	12	0	0	0	36

Fig. 8.4 Example of statistics for threat scenarios

are also useful for quickly showing where the critical threat scenarios are and, on the other hand, where the lower-risk threat scenarios are. For example, if the company's cybersecurity policy performs risk acceptance for risk levels equal to or less than two, it can easily show which risks are greater than two and need to be addressed.

8.2.2 Damage Scenarios

Similar to the statistics for threat scenarios, a statistics chart can be created for damage scenarios. An example is shown in Fig. 8.5. The chart can quickly show which damage scenarios are most likely to occur in the TARA, and the risk levels associated with each damage scenario.

8.2.3 Assets

An interesting statistical analysis is the distribution of risk across assets. An example is shown in Fig. 8.6. In this diagram, we can see which assets have the higher risks associated with them. This information can be useful in the further engineering process. For example, the higher-risk assets can be prioritized when defining protection controls. Observing the results shown in Fig. 8.6, the cybersecurity team might suggest to the project management to prioritize the cybersecurity design for the assets *bootloader*, *CAN*, *software updates* and *OEM applications*, for example.

Statistics for damage scenarios							
Damage scenario	Count occurrences *Asset-Threat scenario-Damage scenario*						
	Total	Occurrences per risk level					
		5	**4**	**3**	**2**	**1**	**Risks > 2**
ECU out of service	1094	324	625	108	30	7	1057
Vehicle out of service, or with degraded funct. [...]	860	435	288	30	102	5	753
Driver distraction, no injuries	678	600	45	33	0	0	678
Disclosure of PII, e.g., location tracking [...]	497	154	112	228	3	0	494
Unavailability of non-safety information [...]	465	0	204	231	20	10	435
Software reverse engineering, product piracy	398	0	89	109	184	16	198
Unauthorized software/data installed in the vehicle	347	198	48	65	36	0	311

Fig. 8.5 Example of statistics for damage scenarios

Statistics for assets												
Asset	Count occurrences *Asset-Threat scenario-Damage scenario*											
	Total	Without controls						With controls				
		5	**4**	**3**	**2**	**1**	**Risks > 2**	**5**	**4**	**3**	**2**	**1**
Bootloader	789	412	163	106	78	30	681	0	0	0	205	584
CAN	657	503	88	33	20	13	624	0	0	0	304	353
Software updates	625	344	178	67	25	11	589	0	0	0	200	425
OEM applications	506	304	128	55	19	0	487	0	0	0	128	378
OBD-II	354	26	104	105	63	56	235	0	0	0	171	183
Private keys and passwords	320	254	46	20	0	0	320	0	0	0	21	299
Ethernet (DoIP)	286	189	70	24	0	3	283	0	0	0	80	206
Ethernet (SOME/IP)	186	102	56	28	0	0	186	0	0	0	65	121
FlexRay	143	77	48	14	4	0	139	0	0	0	88	55
JTAG	130	12	89	18	11	0	119	0	0	0	30	100
Payment information	105	15	66	20	4	0	101	0	0	0	5	100
Account credentials	103	44	22	22	10	5	88	0	0	0	22	81
User profile data	103	45	40	18	0	0	103	0	0	0	23	80
Calibration data	32	0	5	20	7	0	25	0	0	0	0	32
Total	4339	2327	1103	550	241	118	3980	0	0	0	1342	2997
Total (%)	100%	54%	25%	13%	6%	3%	100%	0%	0%	0%	34%	75%

Fig. 8.6 Example of statistics for assets

In addition, the diagram becomes even more interesting when statistics are performed after cybersecurity controls are mapped to risks and a risk evaluation is repeated. As can be seen in Fig. 8.6, the risk levels on the right side of the figure show the residual risk levels after applying cybersecurity controls.

8.3 Practical Case Studies

In this chapter, we continue with the practical case studies developed throughout the book, using the damage scenarios defined in Chap. 6 and the threat scenarios defined in Chap. 7. We apply the risk matrix in Table 8.1 to calculate the risk values.

8.3.1 Case Study 1: Automotive Embedded Device

We evaluate the two threat scenarios for the asset *application software* that we identified in Sect. 7.4.1, namely the threat scenarios *12.3 The software is manipulated before the update process (and is therefore corrupted), although the update process is intact* and *Bypass software signature verification by side channel attack*. We apply the damage scenarios we evaluated in Table 6.12 and 6.13 in Sect. 6.5.1. For example, the damage scenarios defined in Table 6.12 represent the complete list of damage scenarios identified for the project. When analyzing the specific threat scenarios for the assets, not every damage scenario applies to every threat scenario. For example, for both of the threat scenarios in this example, we excluded the damage scenario *Disclosure of personal data*. The risk evaluation for both threat scenarios and for the different stakeholder views (i.e., road user, OEM) are provided in Table 8.5, 8.6, 8.7, and 8.8.

8.3.2 Case Study 2: Vehicle-Backend Architecture

In our second case study, vehicle backend architecture, we selected the asset *Sensitive data in transit* and identified and evaluated the threat scenario *6.2 Man in the middle attack/session hijacking* in Sect. 7.4.2. The damage scenarios have been described for three different stakeholders, i.e., road user, OEM and 3rd party provider, in Table 6.16, 6.17 and 6.18 respectively. We evaluate the risk for each stakeholder view and show the results in Table 8.9, 8.10, and 8.11, respectively.

8.3.3 Case Study 3: Automotive Manufacturing Plant

In this case study, the threat scenario *System Firmware (T0857)* affecting the asset *software updates* of a manufacturing plant is evaluated for different damage scenarios for the stakeholders road user and supplier. The damage scenarios are evaluated in Table 6.18 and 6.19. The threat scenario and its attack path are analyzed in detail in Sect. 7.4.3. The results of the risk evaluation for both stakeholders are presented in Table 8.12 and 8.13.

Table 8.5 Risk evaluation of the road user impact for our Case Study 1

Asset: Application Software

Threat scenario: 12.3 The software is manipulated before the update process (and is therefore corrupted), although the update process is intact

Attack feasibility rating: Medium

Damage Scenario [WP-15-01]	**Risk value for each category [WP-15-07]**				**Maximal risk value**
	Road User View				
	Safety	**Financial**	**Operational**	**Privacy**	**[WP-15-07]**
Vehicle or component (e.g., ECU) gets damaged	1	2	4	1	4
Sensors/warning systems provide wrong information	3	3	3	1	3
Unavailability of non-safety information (e.g., traffic information, music, address book)	1	2	4	1	4
Publication of vulnerability	1	1	1	1	1

Table 8.6 Risk evaluation of the OEM impact for our Case Study 1

Asset: Application Software

Threat scenario: 12.3 The software is manipulated before the update process (and is therefore corrupted), although the update process is intact

Attack feasibility rating: Medium

Damage Scenario [WP-15-01]	**Risk value for each category [WP-15-07]**				**Maximal risk value [WP-15-07]**
	OEM View				
	Intellectual Property	**Image/Reputation**	**Financial**	**Operational**	
Vehicle or component (e.g., ECU) gets damaged	1	2	2	1	2
Sensors/warning systems provide wrong information	1	3	2	1	3
Unavailability of non-safety information (e.g., traffic information, music, address book)	1	2	2	1	2
Publication of vulnerability	2	3	2	2	3

Table 8.7 Risk evaluation of the road user impact for our Case Study 1

Asset: Application Software

Threat scenario: Bypass software signature verification by side channel attack

Attack feasibility rating: Very low

Damage Scenario [WP-15-01]	Risk value for each category [WP-15-07]				Maximal risk value [WP-15-07]
	Road User View				
	Safety	Financial	Operational	Privacy	
Vehicle or component (e.g., ECU) gets damaged	1	1	2	1	2
Sensors/warning systems provide wrong information	1	1	1	1	1
Unavailability of non-safety information (e.g., traffic information, music, address book)	1	1	2	1	2
Publication of vulnerability	1	1	1	1	1

Table 8.8 Risk evaluation of the OEM impact for our Case Study 1

Asset: Application Software

Threat scenario: Bypass software signature verification by side channel attack

Attack feasibility rating: Very low

Damage Scenario [WP-15-01]	Risk value for each category [WP-15-07]				Maximal risk value [WP-15-07]
	OEM View				
	Intellectual Property	Image/ Reputation	Financial	Operational	
Vehicle or component (e.g., ECU) gets damaged	1	1	1	1	1
Sensors/warning systems provide wrong information	1	1	1	1	1
Unavailability of non-safety information (e.g., traffic information, music, address book)	1	1	1	1	1
Publication of vulnerability	1	1	1	1	1

Table 8.9 Risk evaluation of the road user impact for our Case Study 2

Asset: Sensitive data in transit

Threat scenario: 6.2 Man in the middle attack/session hijacking

Attack feasibility rating: Medium

Damage Scenario [WP-15-01]	Risk value for each category [WP-15-07]				Maximal risk value [WP-15-07]
	Road User View				
	Safety	Financial	Operational	Privacy	
Disclosure of business relevant data	1	1	1	1	1
Using services without valid licenses	1	2	3	1	3

Table 8.10 Risk evaluation of the OEM impact for our Case Study 2

Asset: Sensitive data in transit

Threat scenario: 6.2 Man in the middle attack/session hijacking

Attack feasibility rating: Medium

Damage Scenario [WP-15-01]	Risk value for each category [WP-15-07]				Maximal risk value [WP-15-07]
	OEM View				
	Intellectual Property	Image/ Reputation	Financial	Operational	
Disclosure of business relevant data	3	3	1	1	3
Using services without valid licenses	1	1	2	1	2

Table 8.11 Risk evaluation of the 3rd party provider impact for our Case Study 2

Asset: Sensitive data in transit

Threat scenario: 6.2 Man in the middle attack/session hijacking

Attack feasibility rating: Medium

Damage Scenario [WP-15-01]	Risk value for each category [WP-15-07]				Maximal risk value [WP-15-07]
	3rd party provider View				
	Intellectual Property	Image/ Reputation	Financial	Operational	
Disclosure of business relevant data	3	2	1	1	3
Using services without valid licenses	1	1	1	1	1

Table 8.12 Risk evaluation of the road user impact for our Case Study 3

Asset: Software updates

Threat scenario: System Firmware (T0857)

Attack feasibility rating: Medium

Damage Scenario [WP-15-01]	Risk value for each category [WP-15-07]				Maximal risk value [WP-15-07]
	Road User View				
	Safety	Financial	Operational	Privacy	
Manipulated software introduced in the final product (ECU)	4	3	4	1	4
Production machines, or local servers infected with ransomware	1	1	1	1	1
Production stop longer than one day	1	1	1	1	1
Manipulated production data, or BOM, lead to defective final product	4	3	4	1	4

Table 8.13 Risk evaluation of the supplier impact for our Case Study 3

Asset: Software updates

Threat scenario: System Firmware (T0857)

Attack feasibility rating: Medium

Damage Scenario [WP-15-01]	Risk value for each category [WP-15-07]				Maximal risk value [WP-15-07]
	Supplier View				
	Intellectual Property	Image/ Reputation	Financial	Operational	
Manipulated software introduced in the final product (ECU)	1	3	2	1	3
Production machines, or local servers infected with ransomware	1	4	3	3	4
Production stop longer than one day	1	4	3	3	4
Manipulated production data, or BOM, lead to defective final product	1	4	4	2	4

Reference

1. ISO/SAE 21434:2021 Road vehicles—Cybersecurity engineering. Standard, International Organization for Standardization/Society of Automotive Engineers, Geneva, CH (2021). https://www.iso.org/standard/70918.html

Chapter 9
Risk Treatment Decision

Abstract Once the risk assessment has been completed, the next step is to decide what actions to take to address the resulting risks. In this step, the cybersecurity team decides, in accordance with ISO/SAE 21434 and corporate policy, whether the risks can be avoided, reduced, shared, or accepted. Each individual risk that is greater than a minimum value must be analyzed. If cybersecurity controls are applied to reduce the risks, the new risk values are calculated. In this step of TARA, cybersecurity claims and cybersecurity goals are defined when risks are accepted or shared, or when cybersecurity controls are applied. We continue with our case studies and provide examples of various risk treatment decisions, including the mapping of cybersecurity controls. Although this is the final formal step of the TARA, we describe how the TARA is actually an evolving document, and the process can be iterated as often as needed.

We have arrived at the final formal step of the TARA process, which is to decide about the identified risks. The decision has to be made for each identified risk, i.e., for each threat scenario. In the literature and in ISO/SAE 21434, there are basically four classical strategies or decisions for risk treatment: *risk avoidance*, *risk reduction*, *risk sharing*, and *risk acceptance*. In addition, there are two key elements for automotive TARA in the context of ISO/SAE 21434 that we address in this step, namely the definition of *cybersecurity goals* and *cybersecurity claims*. The step risk treatment decision in the overall TARA process is depicted in Fig. 9.1.

Necessary input for this chapter:

- Risk evaluation of each threat scenario

The outcome of this chapter is:

- Risk treatment decision
- Cybersecurity controls
- Cybersecurity goals
- Cybersecurity claims

R. do Carmo and A. Schlensog, *Automotive Threat Analysis and Risk Assessment in Practice*, https://doi.org/10.1007/978-3-662-69614-9_9

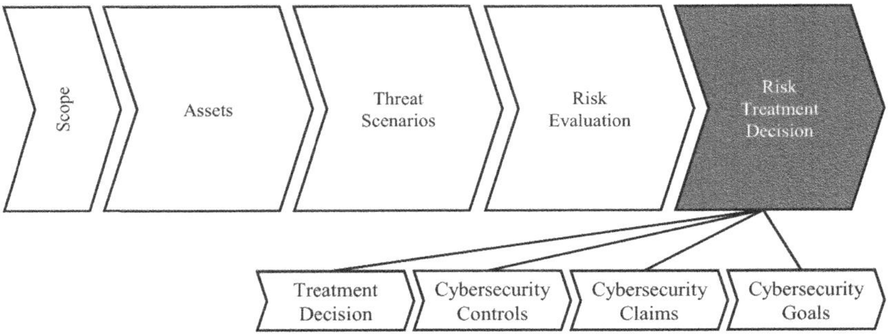

Fig. 9.1 The risk treatment decision in the overall TARA process

9.1 Risk Treatment Decision and Risk Tolerance

The purpose of creating a TARA is to identify the risks of an item and to do something about those risks. Typically, the TARA project will make decisions based on the organization's predefined risk management policies. ISO/SAE 21434 specifies that for threat scenarios with a risk value of one, these risks do not need to be handled according to the standard (see *[PM-06-08]* in [1]). Only when these risks have cybersecurity implications do they need to be addressed, but with less rigor than in the standard. For risk values greater than one, a risk treatment decision must be made.

9.2 Risk Avoidance

Risk avoidance consists of changing the design of the item to eliminate the source of the risk. For example, if there is a risk that the software of an ECU can be manipulated by a debugging protocol running on the JTAG interface, the risk can be eliminated by removing the JTAG interface. Another option would be to leave the JTAG interface on the hardware design, but remove the debugging protocol from the software build. The risk avoidance approach can be useful in the concept phase of the item, for example, when performing an initial TARA. In the concept phase, it is possible to make high-level decisions about the SW or HW design and make architectural changes without incurring high costs.

9.3 Risk Reduction

Risk is a combination of impact and likelihood (feasibility). Theoretically, reducing risk could mean either reducing impact or reducing likelihood. Reducing the impact of a threat scenario does not make sense because the impact (in TARA, modeled by damage scenarios) does not depend on the threat scenario. Therefore, risks can be reduced by reducing their likelihood, i.e., by increasing their feasibility. In practice, increasing the feasibility of an attack means making the threat scenario more difficult to execute.

The reduction of risks is achieved by technical countermeasures for the item, called *cybersecurity controls*. In addition to assigning cybersecurity controls, *cybersecurity goals* must be specified when deciding on risk reduction.

Cybersecurity controls and concept design
The assignment of cybersecurity controls in a TARA project typically has the primary purpose of providing a recommendation from the cybersecurity team to the system design to mitigate the identified risks. The cybersecurity controls are usually concept-level mitigations from standard or corporate-specific catalogs, such as *secure boot*, *secure update*, *firewall*, and others. The TARA is a living document, and the cybersecurity controls proposed in the TARA must be refined during the design and engineering process. Cybersecurity requirements, taking into account the specific item and its operational environment, must be assigned to the cybersecurity controls. The completeness, correctness, and consistency of the cybersecurity controls regarding the cybersecurity goals and cybersecurity claims must be verified. This further analysis of the cybersecurity controls goes beyond the TARA process and is part of the design of the item's cybersecurity concept (see *Sect. 9.5 Cybersecurity Concept* in [1]).

9.3.1 Cybersecurity Goals

Cybersecurity goals are concept-level cybersecurity requirements associated with one or more threat scenarios. Cybersecurity goals describe the requirements for protecting assets against threat scenarios. Particularly in the concept phase, cybersecurity goals help understand the decision to apply cybersecurity controls to the item and make architectural design decisions. Cybersecurity goals are also important for making supplier capability decisions (see *Sect. 7.4.1* in [1]). In a request for quotation for the development of an item, the OEM or supplier needs to specify the cybersecurity goals that are relevant and must be met by their candidate supplier. Examples of cybersecurity goals may include:

- PII stored in the item shall be protected from unauthorized access.
- The integrity and authenticity of sensitive messages shall be ensured.
- Diagnostic access shall be protected from unauthorized access.
- Certificates shall not be modified by unauthorized users.

9.3.2 *Cybersecurity Controls*

Cybersecurity controls are usually technical mitigations to prevent the threat scenario. Procedural countermeasures can also be applied as cybersecurity controls. When we discussed the threat identification process, we emphasized that threat identification is a creative process that requires the cybersecurity team to dive deep into the technical details of the item and think like an attacker. For this reason, we recommend conducting the threat identification process through brainstorming sessions, rather than relying solely on standard catalogs. On the other hand, when assigning cybersecurity controls, we recommend using standard catalogs. There are many cybersecurity controls already available in the industry, including well-known conceptual approaches such as secure boot, so the chances of finding cybersecurity controls for the threat scenario are quite high. Following the well-known layered approach to cybersecurity, multiple layers of defense are typically required to mitigate risk. Several cybersecurity controls from standard catalogs can be applied to the threat scenarios. Using known cybersecurity controls also has the advantage of facilitating the design process of the item.

There are many sources for cybersecurity controls. The sources we described earlier as threat catalogs also contain cybersecurity controls, and in most cases they are interrelated. For example, the UN Regulation No. 155 [2] has a list of mitigations that correspond to the threats listed in its *Annex 5*. The MITRE ATT&CK® [3] framework is also a highly recommended source of mitigations. The framework presents several mitigations for the threats (techniques) described. OWASP® Top 10 [4] provides an exhaustive list of countermeasures to prevent the described threats. We encourage the analysis of the OWASP® Top 10 for all types of items (i.e., embedded or IT/OT systems), as many of them are basic mitigations that can be universally applied (e.g., see the mitigations of *A02:2021—Cryptographic Failures* in [4]). Cybersecurity controls are also available in some international standards, such as the ISO/IEC 27000-series [5], IEC 62443-4-2 [6], or NIST SP 800-53 [7]. Additional cybersecurity controls are available in the ENISA Threat Landscape (ETL) [8]. Specifically for automotive systems, cybersecurity controls are well described in [9, 10].

9.3.2.1 Categorization of Cybersecurity Controls

There are various approaches to a taxonomy of cybersecurity controls. We do not address a control taxonomy in this book because it is rather beyond the boundaries of the TARA process. In short, cybersecurity controls can be classically divided into *preventive*, *detective*, and *corrective controls*. Preventive controls aim to prevent cybersecurity attacks from occurring in the first place. Most of the technical cybersecurity controls at TARAs are preventive. Detective controls act when the cybersecurity attack has been successfully executed. Corrective controls take action after a successful cybersecurity attack. Other classification schemes are based on specific security attributes of the controls, such as physical and logical controls,

controls related to access control, training, configuration management, personal data processing, and others.

The cybersecurity team and the organization are free to use or design a classification scheme. For the TARA process, it is necessary to link the cybersecurity controls to the specific threat scenarios where risk needs to be mitigated. Since the TARA is modeled by cybersecurity properties, it is important to understand and document how the selected cybersecurity controls relate to the cybersecurity property of the asset at risk from the threat scenario. As a simple example, a cybersecurity control that only protects the cybersecurity property of *confidentiality* should not be applied to a threat scenario that only compromises the *availability* of the asset. For this reason, when the cybersecurity team creates its own catalogs, it can be very useful to document the affected cybersecurity property for each control in the catalog. We provide some examples of typical controls for automotive embedded systems in Table 9.1 and for IT/OT systems in Table 9.2.

Table 9.1 Typical cybersecurity controls for automotive embedded systems, including a sample mapping to the corresponding cybersecurity properties

Cybersecurity control	**Related to cybersecurity property**[a]					
	C	**I**	**Av**	**Au**	**NR**	**Acc**
Secure boot		X	X	X		
Secure flashing (secure programming)	X	X		X	X	X
Message authentication		X		X	X	X
Secure onboard communication (SecOC)		X				
Secure update	X	X		X	X	X
Secure access diagnostics				X		
Secure life cycle management[b]	n/a	n/a	n/a	n/a	n/a	n/a
Protection/Deactivation of debug interfaces	X	X	X	X		
Secure storage of data	X	X		X		
Encryption of data in transit	X	X				
Networks separation	X	X	X			
Hardware security module (HSM)	X	X	X	X		
Firewall		X	X			
Intrusion detection (and prevention) system		X	X			
Protection against voltage glitching attacks		X	X			
Redundant communication systems			X			
Plausibility checks from different sources		X	X			

[a]C: confidentiality, I: integrity, Av: availability, Au: authenticity, NR: non-reputation, Acc: accountability

[b]Some cybersecurity controls cannot be directly related to a cybersecurity property

Table 9.2 Typical cybersecurity controls for IT/OT systems, including a sample mapping to the corresponding cybersecurity properties

Cybersecurity control	Related to cybersecurity property[a]					
	C	I	Av	Au	NR	Acc
Access control	X			X	X	X
Secure storage of sensitive data	X	X				
Encryption of data in transit (e.g., using TLS)	X	X				
Verification of cryptographic algorithms, configurations, and functions[b]	n/a	n/a	n/a	n/a	n/a	n/a
Server-side input validation	X	X	X			
Secure path management process	X	X	X			
Physical tampering detection		X				
Hardware security module (HSM)	X	X	X	X		
Firewall		X	X			
Intrusion detection (and prevention) system		X	X			
Emergency power			X			
Authorization enforcement and least privilege	X	X		X	X	X
Minimization of Wi-Fi propagation	X	X	X	X		
Multifactor authentication				X		
Data backup			X			
Use of cryptography	X	X		X	X	X
Segregation of networks	X	X	X			
Logging[b]	n/a	n/a	n/a	n/a	n/a	n/a

[a]C: confidentiality, I: integrity, Av: availability, Au: authenticity, NR: non-reputation, Acc: accountability
[b]Some cybersecurity controls cannot be directly related to a cybersecurity property

9.3.2.2 Attack Feasibility Rating

The cybersecurity controls shall reduce the attack feasibility rating for the threat scenario for which the cybersecurity control is requested. The methodology for evaluating the attack feasibility rating is the same as described in Sect. 7.3. If the attack potential-based approach was used to evaluate the original threat scenario, the cybersecurity controls will typically increase the rating of at least one of the categories *elapsed time*, *specialist expertise*, *knowledge of the item or component*, *window of opportunity*, and *equipment*.

Residual risk may remain *high*
Residual risk, also known as *mitigated risk*, refers to the remaining level of risk associated with an asset following the implementation of controls. It is supposed to be the risk that is present when the attacker tries to perform the

attack path even when the technical control is in place. In simple words, the attacker trying to break or circumvent the control. The magnitude of residual risk can vary depending on the risk matrix employed (assuming the use of a matrix). It is possible for the residual risk to remain higher than one in certain situations. For instance, if we consider the risk matrix presented in Table 8.1 of Sect. 8.1.1, the minimum residual risk achievable is two. This occurs when the impact rating for the affected damage scenario is classified as *severe*. Some controls, however, might turn the attack path extremely difficult to perform, just like combining many technical controls, or demanding to break state-of-the-art cryptography. In such cases, the cybersecurity team can choose to accept the residual risk by documenting a cybersecurity claim supported by a well-justified rationale. If the control virtually makes the vulnerability used in the selected attack path not existent anymore, it is reasonable not to follow the attack path anymore, which makes the threat scenario not applicable. This case must be well documented as cybersecurity claim, highlighting the impact of the control on the threat scenario.

9.4 Risk Sharing and Risk Acceptance

Risk sharing is the sharing of risk with other stakeholders through contracts (e.g., license agreements, terms and conditions) or risk transfer through insurance. A rationale (cybersecurity claim) for risk sharing must be provided.

In situations where the level of risk is low, a reasonable decision is to accept the risk. The risk level is low in situations where the impact or likelihood of the threat scenario actually occurring (attack feasibility rating) is low. Accepting the risk does not mean completely ignoring the risk. A rationale (cybersecurity claim) must be provided to justify the acceptance of the risk. If the TARA is re-evaluated after a new vulnerability is discovered, the accepted risks may become a higher risk value and can no longer be accepted. For this reason, accepted risks shall be documented in the TARA and shall not be removed.

9.4.1 Cybersecurity Claims

In the context of ISO/SAE 21434, a cybersecurity claim is defined as a statement about risk [1]. These claims serve as a rationale, explaining the reasoning behind decisions to accept or share risk. They play a crucial role in cybersecurity monitoring, as they need to be reassessed whenever new vulnerabilities are discovered. This is because the emergence of new vulnerabilities can alter risk values, potentially rendering previous claims for accepting or sharing risk no longer acceptable.

Consider, for example, a threat scenario with a risk level of three in the operational category for the road user (assuming other categories have lower risk levels). In this case, the OEM might decide to share the risk with an insurance company, which agrees to reimburse the road user in the event of an incident. The cybersecurity claim in this scenario might explain that the risk level is set at three due to the high impact (i.e., high cost to the road user), but the probability of the threat scenario occurring is low. Under this premise, the insurance company accepts the contract with the OEM. However, if a new vulnerability is discovered that raises the risk level to four or five, the existing contract between the OEM and the insurance company would no longer be valid. This is because the insurance company had only agreed to cover incidents up to a risk level of three. In such a situation, a new risk treatment decision would be required, necessitating a reassessment of the cybersecurity claim and potentially renegotiating the terms of the risk-sharing agreement with the insurance company.

9.5 After TARA is Before TARA—Risk Re-Evaluation

The final step in a classic TARA process is to perform a new risk evaluation, after cybersecurity controls have been applied to the threat scenarios. This is not to say that the process ends here. In fact, the TARA *risk evaluation-risk treatment* loop can be repeated as many times as needed, making the term "last step of a TARA" somewhat misleading. A TARA is intended to be a living document that evolves throughout the product life cycle. This means that the risk evaluation process is not a one-time event, but rather a continuous process that must be revisited whenever new vulnerabilities are discovered, or when significant changes are made to the system or its environment. Assuming that we are working on a first TARA for a particular project, the risk re-evaluation consists of performing the risk evaluation considering the new attack feasibility rating for the threat scenarios to which cybersecurity controls are applied. This re-evaluation is crucial to ensure that the implemented controls are effective and that the risk has been reduced to an acceptable level. Although there is no specific term for risk re-evaluation in the context of ISO/SAE 21434, in other frameworks or standards the new risk is usually referred to as *residual risk* or *mitigated risk*. These terms refer to the risk that remains after the risk treatment measures have been applied.

9.5.1 Statistics

Similar to the statistical analysis presented for the assets, threat scenarios, and damage scenarios in Sect. 8.2, we can perform statistics on the cybersecurity controls. We present an example of statistics for cybersecurity controls in Fig. 9.2.

Statistics for cybersecurity controls									
	Count occurrences of the cybersecurity control for risks								
Control	Total	Occurrences per risk level						Control Count Percentage	Control Count Cumulative
		5	4	3	2	1	Risks > 2		
Hardware security module (HSM)	1787	965	563	120	119	20	1648	41,18%	41,18%
Secure boot	678	410	210	34	24	0	654	15,63%	56,81%
Secure storage of data	367	16	81	250	20	0	347	8,46%	65,27%
Intrusion detection (and prevention) system	348	0	128	180	40	0	308	8,02%	73,29%
Secure update	248	100	48	65	35	0	213	5,72%	79,00%
Secure onboard communication (SecOC)	230	68	86	13	63	0	167	5,30%	84,31%
Secure access diagnostics	203	43	88	40	22	10	171	4,68%	88,98%
Firewall	134	23	87	10	6	8	120	3,09%	92,07%
Protection/Deactivation of debug interfaces	124	32	45	40	7	0	117	2,86%	94,93%
Secure flashing (secure programming)	123	26	40	22	35	0	88	2,83%	97,76%
Networks separation	63	20	23	20	0	0	63	1,45%	99,22%
Encryption of data in transit	34	8	12	10	4	0	30	0,78%	100,00%

Fig. 9.2 Example of statistics for cybersecurity controls

We can observe that the number of times a cybersecurity control has been mapped to the threat scenarios can be an interesting indicator when discussing the efficiency or importance of cybersecurity controls. If we look at the cumulative value in the last column, we can see that the first 5 controls cover 80% of all controls, or 80% of all risks. We emphasize that the cumulative value, and in general, the statistics on cybersecurity controls, should not be misunderstood and lead to wrong assumptions. All cybersecurity controls are mapped out for a reason. If we look at the remaining 20% of the cumulative value, there are still many high-risk levels covered by the cybersecurity controls in the 20%. Certainly, there are good reasons for using this statistic. For example, if we look at the last control in Fig. 9.2, *Encryption of data in transit*, the management level might notice that there is a lower number of risks (34) associated with this control. In this example, a valid discussion can be started assuming that encryption of data in transit is too expensive for the particular project. Management, together with the cybersecurity team, can analyze whether the risks could be minimized in another way, for example, by applying other cybersecurity controls that also match the threat scenarios, or by changing the design of the item.

9.6 Practical Case Studies

We continue with our three practical case studies and perform the risk treatment decision. We present examples of the treatment decision and the re-evaluation of the adoption of certain cybersecurity controls. In the example of statistics for assets in Fig. 8.6, we already gave a glimpse of the controls. We showed that it can be quite convenient to show how the risk level associated with each asset decreases after applying cybersecurity controls.

9.6.1 Case Study 1: Automotive Embedded Device

For our first case study, we have described two threat scenarios, namely *12.3 The software is manipulated before the update process (and is therefore corrupted), although the update process is intact* and *Bypass software signature verification by side channel attack*. We have calculated the risk value for the example asset *Application Software* in Sect. 8.3.1. The first threat scenario has an attack feasibility rating of *medium*. Since there are risk values greater than one associated with this threat scenario, we can apply controls to reduce the attack feasibility rating and ultimately the resulting risk values. The second threat scenario *Bypass software signature verification by side channel attack*, on the other hand, already has an attack feasibility rating value of *very low*. This is the lowest value we can achieve, and therefore we cannot apply cybersecurity controls to it. We take a closer look at the resulting risk values for this threat scenario and make a risk treatment decision.

In Table 9.4, we present the risk evaluation for the stakeholder road user that we have already calculated in Sect. 8.3.3. We include the risk treatment decision for each damage scenario. Our company policy requires us to mitigate risks greater than two. Risk level two can be accepted under certain conditions (e.g., if there is no safety impact). Since there are risk values greater than two for some damage scenarios, we decided to apply controls to the threat scenario. Suppose we apply some controls listed in Tables 9.1 and 9.2. The cybersecurity controls result in a new feasibility rating for the attack path. We show the new feasibility evaluation in Table 9.3. The highlighted values represent the rating values that have changed due to the cybersecurity controls (see Table 7.9). We perform the new risk evaluation by combining the impact rating for the damage scenarios described in Table 6.12 with the new attack feasibility rating. The results are shown in Table 9.4. We have excluded the calculations of the risk values per impact category from Table 9.4 due to space limitations. The calculation of the *new risk* after applying the cybersecurity controls is the maximum value among all involved rating categories (i.e., safety, privacy, operational, and financial).

We provide examples for cybersecurity goals and cybersecurity claims after applying the risk treatment decision at both threat scenarios:

- *Cybersecurity goal:* Software updates shall be protected against unauthorized manipulation before reaching the vehicle.
- *Cybersecurity goal:* Debugging interfaces shall be protected against unauthorized access.
- *Cybersecurity goal:* The ECU shall implement countermeasures against power glitching attacks.
- *Cybersecurity goal:* The ECU shall verify the authenticity of the software at boot process.
- *Cybersecurity claim:* The risk value two concerning the damage scenario *Vehicle or component (e.g., ECU) gets damaged* (Table 9.4) is accepted because the reason for the risk value two is due to the impact category operational. The safety impact category has a resulting risk value of one.

Table 9.3 New attack feasibility rating for threat scenario *12.3 The software is manipulated before the update process (and is therefore corrupted), although the update process is intact* after applying cybersecurity controls

Steps of the attack path	Elapsed time	Specialist expertise	Knowledge	Window of oppor.	Equipment	Attack feasib.
The attacker compromises the back-end server of the OEM/Tier (e.g., insider attacker)	**≤6 months**	Proficient	**Confid.**	Unlimited	Standard	**Very low**
Manipulated software update is uploaded to the back-end server. The software does not have any cryptographic signature	≤6 months	Expert	**Confid.**	Easy	**Bespoke**	**Very low**
The manipulated software is installed on the vehicle by normal update process (no cryptographic signature verification)	≤1 day	Layman	Public	Unlimited	Standard	High
Overall rating for the attack path	**≤6 months**	**Expert**	**Confid.**	**Easy**	**Bespoke**	**Very low**

- *Cybersecurity claim:* The risk value two concerning the damage scenario *Unavailability of non-safety information* (Table 9.4) is accepted because no safety risk is implicated.
- *Cybersecurity claim:* The risk value two concerning the damage scenario *Vehicle or component (e.g., ECU) gets damaged* (Table 8.7) is accepted because the reason for the risk value two is due to the impact category operational. The safety impact category has a resulting risk value of one.
- *Cybersecurity claim:* The risk value two concerning the damage scenario *Unavailability of non-safety information* (Table 8.7) is accepted because no safety risk is implicated.

9.6.2 Case Study 2: Vehicle-Backend Architecture

For our case study about a connected car architecture, we performed the same process of risk treatment decision and risk re-evaluation as we did for our first case study.

Table 9.4 Risk re-evaluation after of the road user impact for our Case Study 1

Asset: Application Software

Threat scenario: 12.3 The software is manipulated before the update process (and is therefore corrupted), although the update process is intact

Attack feasibility rating (after applying controls): Medium

Damage Scenario [WP-15-01]	Maximal risk value [WP-15-07]	Risk treat. decision [WP-15-08]	New maximal risk value [WP-15-07]
Vehicle or component (e.g., ECU) gets damaged	4	Reduction	2
Sensors/warning systems provide wrong information	3	Reduction	1
Unavailability of non-safety information (e.g., traffic information, music, address book)	4	Reduction	2
Publication of vulnerability	1	Acceptance	1

Since our goal is to explain the process of TARA, we skip some steps and present the results directly. We show the analysis for the risk values related to the stakeholder *OEM*. The initial risk evaluation is provided in Table 8.10.

The results of the risk treatment decision are presented in Table 9.5. Assuming we applied cybersecurity controls to the item and repeated the risk feasibility evaluation process for the affected threat scenario, we present the new maximum risk value calculation in Table 9.5. In this example, there are no cybersecurity claims because all risks have been reduced to the minimum level. The derived cybersecurity goals are as follows:

Table 9.5 Risk re-evaluation after of the OEM impact for our Case Study 2

Asset: Sensitive data in transit

Threat scenario: 6.2 Man in the middle attack/session hijacking

Attack feasibility rating (after applying controls): Very low

Damage Scenario [WP-15-01]	Maximal risk value [WP-15-07]	Risk treat. decision [WP-15-08]	New maximal risk value [WP-15-07]
Disclosure of business relevant data	3	Reduction	1
Using services without valid licenses	2	Reduction	1

- *Cybersecurity goal:* Data in transit shall be protected (by encryption) against unauthorized access.
- *Cybersecurity goal:* Business relevant data shall be identified, and access control mechanisms shall be applied.

9.6.3 Case Study 3: Automotive Manufacturing Plant

The risk values for our third case study have already been presented in Table 8.12 for the road user and Table 8.13 for the supplier. We follow the same approach as in our case studies one and two, making the risk treatment decision for all risk values and applying cybersecurity controls to the risks that need to be mitigated. We present the results of these steps in Tables 9.6 and 9.7 for the road user and the supplier, respectively.

As can be observed in Tables 9.6 and 9.7, not all risks can be reduced to a minimum. Therefore, the cybersecurity team must analyze each risk individually and make a new risk treatment decision. In the previous case studies, we simply accepted the remaining risk of two. We created a cybersecurity claim and argued that there are no safety implications with this risk level, and therefore the organization decides to accept the remaining risks. In this case study, we have the challenge that the risk level two related to the damage scenario *Manipulated software introduced in the final product (ECU)* has the level two because of the category safety for the road user. For this particular case, the cybersecurity team might have to either try

Table 9.6 Risk re-evaluation after of the OEM impact for our Case Study 3

Asset: Software updates

Threat scenario: System Firmware (T0857)

Attack feasibility rating (after applying controls): Very low

Damage Scenario [WP-15-01]	Maximal risk value [WP-15-07]	Risk treat. decision [WP-15-08]	New maximal risk value [WP-15-07]
Manipulated software introduced in the final product (ECU)	4	Reduction	2
Production machines, or local servers infected with ransomware	1	Acceptance	1
Production stop longer than one day	1	Acceptance	1
Manipulated production data, or BOM, lead to defective final product	4	Reduction	1

Table 9.7 Risk re-evaluation after of the supplier impact for our Case Study 3

Asset: Software updates

Threat scenario: System Firmware (T0857)

Attack feasibility rating (after applying controls): Very low

Damage Scenario [WP-15-01]	Maximal risk value [WP-15-07]	Risk treat. decision [WP-15-08]	New maximal risk value [WP-15-07]
Manipulated software introduced in the final product (ECU)	3	Reduction	1
Production machines, or local servers infected with ransomware	4	Reduction	2
Production stop longer than one day	4	Reduction	2
Manipulated production data, or BOM, lead to defective final product	4	Reduction	2

to avoid the risk by eliminating the source of the risk (e.g., not providing software updates for the production machines), or further analyze with the engineering and safety team if the threat scenario has been analyzed correctly. It may be that due to other technical controls, system design, and safety features, the damage scenario *Manipulated software introduced in the final product (ECU)* was not possible at all.

- *Cybersecurity goal:* It shall not be possible to introduce manipulated software into the production machinery.
- *Cybersecurity goal:* Firmware, including updates, has to be developed in a trusted environment, and it shall not be possible to manipulate the firmware during its development.
- *Cybersecurity claim:* Regarding the risk related to the damage scenarios *Production machines, or local servers infected with ransomware*, *Production stop longer than one day*, and *Manipulated production data, or BOM, lead to defective final product* for the supplier risk analysis, the supplier decides to accept the risk level two because the categories image and financial are affected with risk level two, and it is a level that the corporate can manage.

References

1. ISO/SAE 21434:2021 Road vehicles—Cybersecurity engineering. Standard, International Organization for Standardization/Society of Automotive Engineers, Geneva, CH (2021). https://www.iso.org/standard/70918.html
2. United Nations: UN Regulation No. 155—Cyber security and cyber security management system E/ECE/TRANS/505/Rev.3/Add.154 (2021). https://unece.org/transport/documents/2021/03/standards/un-regulation-no-155-cyber-security-and-cyber-security
3. The MITRE Corporation: MITRE ATT&CK®. https://attack.mitre.org/

4. OWASP Top Ten—OWASP Foundation. https://owasp.org/www-project-top-ten/
5. ISO/IEC 27002:2022 Information security, cybersecurity and privacy protection—Information security controls. Standard, International Organization for Standardization/International Electrotechnical Commission (2022). https://www.iso.org/standard/75652.html
6. IEC 62443-4-2:2019 Security for industrial automation and control systems—Part 4-2: Technical security requirements for IACS components. Standard, International Electrotechnical Commission (2019). https://webstore.iec.ch/publication/34421
7. NIST: NIST Special Publication 800-37 Risk Management Framework for Information Systems and Organizations—A System Life Cycle Approach for Security and Privacy (2018). https://doi.org/10.6028/NIST.SP.800-37r2
8. European Union Agency for Cybersecurity (ENISA): ENISA Threat Landscape 2022 (2022). https://www.enisa.europa.eu/publications/enisa-threat-landscape-2022
9. European Union Agency for Cybersecurity (ENISA): ENISA good practices for security of Smart Cars (2019). https://www.enisa.europa.eu/publications/smart-cars
10. Wurm, M.: Automotive Cybersecurity: Security-Bausteine für Automotive Embedded Systeme. Springer Berlin Heidelberg, Berlin, Heidelberg (2022). https://doi.org/10.1007/978-3-662-64228-3. https://link.springer.com/10.1007/978-3-662-64228-3

Appendix

A.1 Asset Catalog

In this section, we present a list of assets that are intended to be used as a generic reference and for example purposes. For this reason, there are assets that can be included in other assets, such as CAN as a generic asset, while the individual protocols running over CAN are separate assets. How the assets are grouped, i.e., the level of abstraction or functional grouping, must be decided by the security team performing the TARA. The format of the table is useful for real-world TARA projects. We have divided the groups into assets for automotive embedded devices and assets for IT/OT systems. The column *Asset [WP-15-02]* describes the name of the asset. The *[WP-15-02]* is a written reference to the corresponding ISO/SAE 21434 work product. This makes it easier to demonstrate conformance to the standard, e.g., for an audit. The last column describes the reason the selected cybersecurity properties are relevant to the asset. This description is useful for documentation purposes. In real TARA projects, we would recommend adding a column called *Comments/Description* to describe any kind of information that is important to know about the assets within a specific project. We have omitted this column for space reasons and because it is rather important in certain projects.

Important note regarding the cybersecurity properties: The evaluation shown in this section is only exemplary, and the cybersecurity properties are selected in a generic manner. While the selection may be appropriate for many real-world projects, the cybersecurity team performing the TARA must evaluate them on a case-by-case basis.

R. do Carmo and A. Schlensog, *Automotive Threat Analysis and Risk Assessment in Practice*, https://doi.org/10.1007/978-3-662-69614-9

A.1.1 Automotive Embedded Devices

See Table A.1

Table A.1 Catalog of assets with cybersecurity properties

Asset [WP-15-02]	Comment/Relevant cybersecurity property for the asset and rationale of the relevance
A. Communication interfaces/protocols	
CAN (Generic CAN bus, messages can be for different purposes)	**Integrity:** while CAN can be used for many purposes, in this example we can assign this property to certain safety-critical messages **Availability:** for the correct operation of the vehicle, the CAN bus availability has to be ensured
UDS	**Integrity:** it shall be ensured that the diagnostics messages are not manipulated **Availability:** UDS services have to be available **Authenticity:** specially for services modifying software and memory areas, it is important to guarantee authenticity
XCP	**Integrity:** it shall be ensured that the diagnostics messages are not manipulated **Authenticity:** similar to UDS, authenticity is relevant when writing memory areas
FlexRay	**Integrity:** similar as for other communication protocols, the integrity of at least some messages has to be guaranteed **Availability:** depending on the usage of the communication protocol, availability has to be guaranteed
LIN	**Integrity:** similar as for other communication protocols, the integrity of at least some messages has to be guaranteed **Availability:** depending on the usage of the communication protocol, availability has to be guaranteed
MOST	**Integrity:** similar as for other communication protocols, the integrity of at least some messages has to be guaranteed **Availability:** depending on the usage of the communication protocol, availability has to be guaranteed
OBD-II	**Integrity:** at least for services used for SW updates over OBD, if any, integrity and authenticity are relevant **Authenticity:** at least for services used for SW updates over OBD, if any, integrity and authenticity are relevant

(Continued)

Table A.1 (Continued)

Asset [WP-15-02]	**Comment/Relevant cybersecurity property for the asset and rationale of the relevance**
	A. Communication interfaces/protocols
Ethernet	**Confidentiality/Integrity/Availability/Authenticity:** the particular protection depends on the service and protocol running over Ethernet. For this reason, protection of confidentiality, integrity, availability, and authenticity in this case is left as an example. Of course, it can be protection mechanisms applied at lower levels, such as MACsec or IPsec
DoIP	**Integrity:** it shall be ensured that, at least certain, diagnostics messages are not manipulated **Availability:** similar to UDS, DoIP services have to be available **Authenticity:** specially for services modifying software and memory areas, it is important to guarantee authenticity
SOME/IP	**Integrity:** pretty depending on the usage, but integrity might be critical for many services **Availability:** due to operational relevance, availability has to be ensured **Authenticity:** depending on the service, authenticity might have to be ensured
JTAG	**Integrity:** it shall be ensured that, at least certain, diagnostics messages are not manipulated **Authenticity:** powerful diagnostics/debugging interface
Wi-Fi	**Confidentiality:** due to its RF nature, sniffing on Wi-Fi is straightforward, and a compromise of further properties possible **Integrity:** similar to other communication protocols, it depends on the service, but integrity most probably needed **Availability:** probably not essential for the vehicle's operation, but might be of relevance for the driver or passenger
Bluetooth	**Confidentiality:** RF nature, ensuring confidentiality can prevent data disclosure and further threats **Integrity:** messages shall not be manipulated **Availability:** relevant for the operation of, e.g., an infotainment system
USB	**Integrity:** messages shall not be manipulated **Availability:** depending on the use case, reliable availability can be expected
eSIM	**Confidentiality:** RF nature, message confidentiality of the utmost importance **Integrity:** relevant to ensure QoS **Availability:** due to connectivity source for infotainment **Authenticity:** authenticity of Internet providers and services has to be ensured

(Continued)

Table A.1 (Continued)

Asset [WP-15-02]	Comment/Relevant cybersecurity property for the asset and rationale of the relevance
	A. Communication interfaces/protocols
Digital/Analog I/O	**Integrity:** mostly used for sensors and actuators, integrity is expected (even if functional safety controls achieve it) **Availability:** due to its operational nature
GPIO, I2C, SPI	**Integrity:** messages shall not be manipulated **Availability:** for operational cases (e.g., interconnection of boards inside an ECU), availability is important
Proprietary Debug Interface	**Integrity:** used for powerful debugging, messages shall not be manipulated **Authenticity:** authenticity of proprietary tools and running protocols have to be guaranteed
	B. Security-relevant data
Public Keys and Certificates	**Integrity:** essential for use **Availability:** different services might rely on public keys and certificates, and they have to be available **Authenticity:** per definition, authenticity has to be ensured (e.g., by relying on PKI)
Private Keys and Passwords	**Confidentiality:** security secrets have to be confidential, per definition **Integrity:** integrity of security secrets is essential for operation **Availability:** availability is required since many services might rely on security secrets **Authenticity:** per definition, authenticity has to be ensured (at least by the initial writing of security secrets)
Account Credentials	**Confidentiality:** credentials for different services must be kept secret **Integrity:** integrity is essential for operation **Availability:** similar to keys, credentials are essentially needed for the services using them **Authenticity:** authenticity has to be ensured (e.g., by relying on PKI)
Unique IDs	**Confidentiality:** pretty much depending on the usage, some unique IDs have to be kept secret (e.g., session IDs) **Integrity:** is essential for the use of the asset **Availability:** as a security artifact, availability has to be ensured for services
Secure Time on Vehicle Bus	**Integrity:** integrity of secure data has to be ensured, i.e., manipulations shall be detectable **Availability:** since many services in the vehicle might depend on real, secure time, availability has to be ensured **Authenticity:** at least the ECUs participating at secure time have to authenticate and trust each other

(Continued)

Table A.1 (Continued)

Asset [WP-15-02]	Comment/Relevant cybersecurity property for the asset and rationale of the relevance
	B. Security-relevant data
Random Numbers	**Integrity:** random numbers usually used for generating hashes and cryptographic operations have to be used in a protected environment to prevent cryptanalysis or attacks on algorithms [55] **Availability:** availability per se is needed for cryptographic services
MAC and Signatures	**Integrity:** integrity of MAC and digital signatures is essential to be ensured **Availability:** because they are used for services, such as varying authenticity of software code, availability is important **Authenticity:** authenticity has to be guaranteed, at least by initial writing in protected areas
Backend Configuration	**Confidentiality:** any important configuration needed for communicating with backend systems might need to be confidential (e.g., for preventing disclosing IP addresses, usernames, etc.) **Integrity:** integrity is essential for operation **Availability:** availability is essential for operation
	C. Personally identifiable information (PII)
Drivers/Passenger Personal Information	**Confidentiality:** PII has to be kept confidential **Integrity:** data integrity is essential for detecting manipulations **Availability:** might depend on the particular application or service relying on the data **Non-repudiation:** applications storing PII could implement non-repudiation mechanisms (e.g., by making the person physically signing a contract) **Accountability:** the system processing data can be related to responsibilities regarding data handling
Payment Information	**Confidentiality:** sensitive data, such as credit card information, has to be confidential **Integrity:** data integrity is important for detecting manipulations **Availability:** might depend on the particular application or service relying on the data **Authenticity:** it shall be ensured that these data come from a trusted source

(Continued)

Table A.1 (Continued)

Asset [WP-15-02]	Comment/Relevant cybersecurity property for the asset and rationale of the relevance
	C. Personally identifiable information (PII)
Geolocation Data	**Confidentiality:** generically, it shall not be possible to link the geolocation to a person; hence, the handling of geolocation data might need to be confidential **Integrity:** for correct operation of services relying on these data **Availability:** might depend on the particular application or service relying on the data **Authenticity:** the source providing geolocation data has to be validated as trusted **Non-repudiation:** non-repudiation might be appropriate for geolocation data provided by external devices **Accountability:** services providing geolocation data can be liable for wrong information
Vehicle Identification Number (VIN)	**Confidentiality:** confidentiality is required to avoid authorized reading of the VIN **Integrity:** VIN shall not be manipulated
Video Data	**Confidentiality:** video sequences can contain PII or other sensitive information and therefore have to be protected against loss of confidentiality **Integrity:** integrity is important depending on the usage **Availability:** services relying on cameras need fully availability **Authenticity:** authenticity of data can be, for example, an important requirement **Non-repudiation:** applications storing PII could implement non-repudiation mechanisms (e.g., by making the person physically signing a contract) **Accountability:** the system processing data can be related to responsibilities regarding data handling
	D. Other data on connected cars
DTC	**Integrity:** diagnostics code shall not be manipulated **Availability:** DTCs have to be available for analyzing failures on the vehicle
Calibration Data	**Integrity:** calibration configuration of ECUs is typically written at production for adjusting the ECU to the vehicle type. Changes by the user are not foreseen, and it has to be protected against unauthorized manipulation **Authenticity:** it can be the case that calibration data is intentionally modified at the workshop, e.g., after a replacement, and the ECU might need to verify the workshop tool to avoid unauthorized changes

(Continued)

Table A.1 (Continued)

Asset [WP-15-02]	Comment/Relevant cybersecurity property for the asset and rationale of the relevance
	D. Other data on connected cars
Logs	**Integrity:** logs can generally provide useful technical information for diagnostics. Therefore, protecting its integrity is needed **Non-repudiation:** in some cases, it can be necessary to guarantee that log files have been written by certain applications, and a non-repudiation mechanism has to be in place
Odometer Counter	**Integrity:** odometer value has to be protected against manipulation
User Profile Data	**Integrity:** user profile data might include some sensitive information for connecting with other services. Therefore, integrity has to be guaranteed **Authenticity:** it has to be ensured that the data come from the user, or another authorized entity
Life cycle State	**Integrity:** ECUs can have a life cycle state for allowing better debugging and development. Once set, it shall not be possible to modify it without authorization
Data on RAM	**Confidentiality:** in extreme cases, e.g., when operating with secret keys, or with PII, it has to be ensured that these data in RAM cannot be accessed without authorization
EDR/DSSAD	**Confidentiality:** while dependent on the legislation, let us assume the strongest requirement. These data have to be protected against unauthorized access **Integrity:** these data are used for after-crash analysis; therefore, integrity has to be guaranteed **Availability:** because data is mostly required by law or by insurance, its availability has to be ensured **Authenticity:** an authenticity protection mechanism can ensure, for example, that these data come from a trusted EDR/DSSAD ECU for the vehicle **Non-repudiation/Accountability:** due to their legal usage, non-repudiation and accountability has to be considered for protection

(Continued)

Table A.1 (Continued)

Asset [WP-15-02]	**Comment/Relevant cybersecurity property for the asset and rationale of the relevance**
	E. Software components
Bootloader	**Integrity:** essential component of the booting process, and of the chain of trust. Manipulations have to be detectable **Availability:** essential for operation, and running the operating system with its applications **Authenticity:** at least at installation time, the source of the software has to be verified to ensure that it is a trusted source
OEM Applications	**Integrity:** the application providing the core functionality of the embedded device shall not be manipulated **Availability:** essential for the main operation of the embedded device **Authenticity:** at least at installation time, the source of the software has to be verified to ensure that it is a trusted source
Supplier Applications	**Integrity:** supplier applications can provide core functionalities and shall not be manipulated **Availability:** essential for the main operation of the embedded device **Authenticity:** at least at installation time, the source of the software has to be verified to ensure that it is a trusted source
3rd Party Applications	**Integrity:** applications from external providers shall not be manipulated. These applications provide extended functionalities, and if compromised, they might get access to core functionalities, and important data of the embedded device **Availability:** depending on the functionality provided, applications from external providers might not be critical. However, it is expected that they are operable when needed **Authenticity:** at least at installation time, the source of the software has to be verified to ensure that it is a trusted source
Software Updates	**Integrity:** software updates shall not be manipulated in order to prevent introducing malware into the embedded device **Availability:** software updates have to be highly available, specially when they introduce security patches **Authenticity:** at least at installation time, the source of the software has to be verified to ensure that it is a trusted source
IP-relevant Data	**Confidentiality:** sensitive parts of the software binary, files, or configuration data, have to be protected against unauthorized access

(Continued)

Table A.1 (Continued)

Asset [WP-15-02]	Comment/Relevant cybersecurity property for the asset and rationale of the relevance
	F. Hardware components
ECU Housing	**Integrity:** the physical embedded device, including its PCB, interfaces, and housing, shall not be manipulated. This is specially relevant in case of critical interfaces that are deactivated by physical means (e.g., removed PINs, bypassed connections) **Availability:** physical component that are critical for the operation have to be highly available
Sensors	**Integrity:** sensors might provide data for comfort functions, and for safety-relevant functions. The integrity of the sensors has to be guaranteed **Availability:** sensors providing critical data have to be highly available
Actuators	**Integrity:** actuators providing critical functionalities have to be protected against manipulations, or manipulations shall be detectable **Availability:** actuators providing critical functionalities are essential for operation
Memory	**Confidentiality:** it might be required to physically hide memory components on the circuit board to make it harder to access them **Integrity:** the integrity of memory blocks is critical for operation **Availability:** memory is an essential component of an embedded device, and has to be available
Power Supply	**Integrity:** the physical integrity of connectors and wires is essential for operation **Availability:** essential for operation

A.1.2 IT/OT Systems

In this section, we present assets that are usually present at manufacturing plants. We generalize the assets by referring to them as *assets for IT/OT systems* because they can be generically applied to any system. However, in this book we put particular focus on the ISO/SAE 21434 for automotive systems, and it is specially interesting, and required by the standard, to address manufacturing plants of vehicles, and components (ECUs) of a vehicle. For this reason, we often refer to manufacturing plants when we give examples for the relevance of the cybersecurity properties (Table A.2).

Table A.2 Catalog of assets with cybersecurity properties

Asset [WP-15-02]	Comment/Relevant cybersecurity property for the asset and rationale of the relevance
Debug/Diagnostics Interfaces	**Integrity:** the integrity of the data exchange between a legitimate client (local or remote) and the diagnostics interface has to be guaranteed to prevent, or detect, manipulations **Availability:** diagnostics services have to be reliable for preventing downtimes in case of troubles **Authenticity:** only trusted entities shall have access to diagnostics services of the production machines
USB	**Integrity:** due to the criticality of the data transmitted (e.g., software updates, production data), manipulations have to be detectable **Authenticity:** accessing to a device over USB has to be performed from authorized personnel only
Wi-Fi	**Confidentiality:** because of the open nature of wireless signals, confidentiality is an essential property for preventing unauthorized access, and several kinds of threats. Even if the data transmitted is per se not confidential, such as production monitoring data, protecting networking packets is necessary **Integrity:** protecting the integrity of the Wi-Fi packets increases the robustness and reliability of the network, and systems relying on this network **Availability:** Wi-Fi networks used in production environments have to be highly reliable **Authenticity:** authenticating communication parties in a wireless network is essential due to its open nature
Fieldbus	**Integrity:** logistics, or production data exchanged between the machines, shall not be manipulated **Availability:** production data exchanged between the production machines is essential and shall have a high reliability
Ethernet	**Integrity:** production data shall not be manipulated in transit **Availability:** it is essential for production and has to be available **Authenticity:** authenticity of the data, or of the network devices joining the network, has to be ensured for preventing threats related to unauthorized access
User Interface (HMI)	**Integrity:** the integrity of an HMI has to be enforced to prevent attacks related to access to systems, and data leakage, among others **Availability:** user interfaces at OT systems are usually used for configuration, and monitoring, and they are therefore essential for operation **Authenticity:** accessing to critical functions within an HMI shall be performed by authorized personnel only

(Continued)

Table A.2 (Continued)

Asset [WP-15-02]	Comment/Relevant cybersecurity property for the asset and rationale of the relevance
Remote Access	**Confidentiality:** especially relevant for data in transit. Remote access for diagnostics purposes is a powerful interface **Integrity:** essential property for preventing, or detecting the manipulation of diagnostics messages in transit **Availability:** the remote access can be used to quickly provide support in case a production machine malfunctions. Therefore, it is important to guarantee high availability **Authenticity:** it shall not be possible that unauthorized entities get remote access to the production plant, and production machinery **Non-repudiation:** in the case of critical services, such as configuration changes, or software updates at the machinery, the entity using the remote access has to be traced, and uniquely identifiable **Accountability:** similar to non-repudiation, a traceability to the entity using the service has to be ensured to prevent accountability
Operating System	**Integrity:** software integrity has to be ensured for preventing executing malware on production machines **Availability:** essential for production **Authenticity:** the operating system itself has to be provided by a trusted source. Administrative functions (e.g., root user for a UNIX system) have to be protected against unauthorized access
Applications	**Integrity:** software integrity has to be ensured to prevent executing malware on production machines **Availability:** essential for production **Authenticity:** the software installed at production machinery has to be authentic. It shall be ensured, at least at installation time, that the software is provided from a trusted source
Databases	**Confidentiality:** databases can contain any kind of sensitive data. Protecting the data against unauthorized access, e.g., at the servers storing data, is of utmost importance **Integrity:** data shall not be modified, or manipulation shall be detected **Availability:** essential property since the operative nature of databases in production plants, and other productive IT/OT environments **Authenticity:** accessing databases has to be performed by authorized entities
Software Updates	**Integrity:** protecting the integrity of software updates is essential for avoiding installing malware on production machines **Authenticity:** software update packages shall be provided from trusted sources

(Continued)

Table A.2 (Continued)

Asset [WP-15-02]	Comment/Relevant cybersecurity property for the asset and rationale of the relevance
Logistics Data	**Integrity:** logistics data exchanged between the production machines is essential **Availability:** logistics data have to be reliable because they are essential for production
Bill of Materials (BOM)	**Integrity:** critical asset for the correct manufacturing of a product. Integrity has to be ensured **Availability:** essential for production and therefore has to be highly available **Authenticity:** it has to be ensured that the BOM received at some part of the manufacturing process has been created by a trusted entity. For example, if the project manager is responsible for building and providing the BOM for production, a digital signature system could be in place
Software Bill of Materials (SBOM)	**Integrity:** the SBOM shall not be tampered with at the manufacturing process. The SBOM is used for dealing with vulnerabilities. An attacker could intentionally modify the software versions listed in the SBOM to make it difficult to provide software patches when a vulnerability is discovered **Authenticity:** the SBOM is used for cybersecurity purposes and, therefore, it shall be authentic
Security-relevant Data	**Confidentiality:** confidentiality must be ensured, as it involves security artifacts, such as private keys for SecOC. **Integrity:** security artifacts are needed for the correct operation of the final product, and therefore, they shall not be manipulated **Availability:** essential part of the final product. Availability has to be ensured **Authenticity:** due to its cybersecurity usage, security artifacts have to come from trusted sources, such as the OEM's PKI
Sensors/Actuators	**Integrity:** sensors and actuators might be present at OT environments for many purposes. Production decisions, and even safety-critical functions, can depend on data exchanged between sensors. Therefore, integrity is a relevant property **Availability:** usually essential for operating. Highly availability has to be ensured **Authenticity:** it can be verified, that the sensors and actuators themselves, and the data they provide, and receive, are trusted
Servers	**Integrity:** data at servers shall not be modified, or manipulation shall be detected **Availability:** servers are usually essential for the operation of an IT/OT system **Authenticity:** accessing servers has to be performed from authorized entities
RFID Cards	**Integrity:** it has to be considered protecting the RFID system against classic RFID integrity threats, such as tags cloning, or replay attacks

(Continued)

Table A.2 (Continued)

Asset [WP-15-02]	Comment/Relevant cybersecurity property for the asset and rationale of the relevance
Firewall Configurations	**Confidentiality:** understanding firewall configurations can help an attacker bypass the firewall. Therefore, access to configuration files has to be protected **Integrity:** a firewall configuration has to be allowed only for authorized administrators **Availability:** essential for the correct and secure operation of a network **Authenticity:** accessing to configuration files has to be allowed only for authorized administrators
Networking Devices	**Integrity:** the software and hardware of networking devices shall not be manipulated **Availability:** networking devices are essential for production, and they have to be available **Authenticity:** enforcing authenticity checks for devices connecting to the network is required to prevent threats related to unauthorized access. Additionally, accessing to the devices, for example, for configuration purposes, has to be executed by trusted entities
Storage Systems	**Confidentiality:** storage systems can contain any kind of sensitive data. Protecting the data against unauthorized access is of utmost importance **Integrity:** data shall not be modified, or manipulation shall be detected **Availability:** essential for operation **Authenticity:** accessing storage systems has to be performed by authorized entities
End-Product/PCB	**Integrity:** in production environments, it has to be ensured that the PCB under production does not get manipulated at some step of the production process
Power Supply	**Integrity:** the physical integrity of connectors and wires is essential for operation **Availability:** essential for operation

A.2 Damage Scenario Catalog

We present a list of generic, though realistic, damage scenarios that can be taken as inspiration for real-world projects. The damage scenarios used in a real-world project will be extremely dependent on the concrete project and the item definition. For this reason, it is not possible to generalize the damage scenarios in such a way that they are universally applicable. However, some damage scenarios, such as ECU out of service, or vehicle leaves trajectory injuring people, are usually shared among several projects. We describe the damage scenarios in Table A.3, and we combine the impact rating for the road user, and for the OEM or supplier in the same table. We abstain from introducing a comment next to the impact rating because the comment only takes major importance in real-world projects. We do the impact rating according to the criteria described in Chap. 6.

Table A.3 Catalog of generic damage scenarios

Damage Scenario [WP-15-01]	Impact Rating [WP-15-04]							
	Road User View				OEM/Supplier View			
	Safety	**Financial**	**Operational**	**Privacy**	**Intellectual Property**	**Image/ Reputation**	**Financial**	**Operational**
Driver distraction, no injuries	Negligible	Negligible	Major	Negligible	Negligible	Negligible	Negligible	Negligible
ECU out of service	Negligible	Moderate	Severe	Negligible	Negligible	Moderate	Moderate	Negligible
Vehicle out of service, or with degraded functionality	Negligible	Major	Severe	Negligible	Negligible	Moderate	Moderate	Negligible
Vehicle leaves trajectory causing injuries	Severe	Major	Severe	Negligible	Negligible	Major	Moderate	Negligible
Vehicle is stolen	Negligible	Major	Severe	Moderate	Negligible	Negligible	Negligible	Negligible
Unavailability of non-safety information (e.g., traffic information, music, address book)	Negligible	Moderate	Severe	Negligible	Negligible	Moderate	Negligible	Negligible
Unauthorized software/data installed in the vehicle	Major	Moderate	Moderate	Negligible	Negligible	Moderate	Moderate	Negligible

(Continued)

Table A.3 (Continued)

Damage Scenario [WP-15-01]	Impact Rating [WP-15-04]							
	Road User View				OEM/Supplier View			
	Safety	Financial	Operational	Privacy	Intellectual Property	Image/ Reputation	Financial	Operational
Fraud (e.g., using a license for a service without having paid for it)	Negligible	Major	Negligible	Negligible	Negligible	Moderate	Moderate	Negligible
Disclosure of PII (e.g., location tracking of one, or more vehicles)	Negligible	Moderate	Negligible	Severe	Negligible	Severe	Major	Negligible
Disclosure of business relevant data	Negligible	Negligible	Negligible	Negligible	Moderate	Moderate	Major	Negligible
Sensors/warning systems provide wrong information	Major	Moderate	Major	Negligible	Negligible	Moderate	Moderate	Negligible
Software reverse engineering, product piracy	Negligible	Negligible	Negligible	Negligible	Severe	Major	Severe	Negligible
Manipulated software introduced in the final product (ECU)	Severe	Major	Severe	Negligible	Negligible	Major	Moderate	Negligible

(Continued)

Table A.3 (Continued)

Damage Scenario [WP-15-01]	Impact Rating [WP-15-04]							
	Road User View				OEM/Supplier View			
	Safety	**Financial**	**Operational**	**Privacy**	**Intellectual Property**	**Image/ Reputation**	**Financial**	**Operational**
Production machines, or local servers infected with ransomware	Negligible	Negligible	Negligible	Negligible	Negligible	Severe	Major	Major
Production stop longer than one day	Negligible	Negligible	Negligible	Negligible	Negligible	Severe	Major	Major
Large-scale data breach including cybersecurity artifacts, and logistics data	Negligible	Negligible	Moderate	Negligible	Major	Severe	Major	Moderate
Manipulated production data, or BOM, lead to defective final product	Severe	Major	Severe	Negligible	Negligible	Severe	Severe	Moderate

A.3 Threat Catalog

We provide a detailed overview of the threat catalog outlined in UN Regulation No. 155. While it is important to note that the catalog does not cover every possible threat, we recommend using it as a solid foundation and then expanding it by incorporating additional threats into the TARA process. The catalog includes diverse threat groups that comprehensively address various aspects of modern connected vehicles, such as communication with back-end servers, V2X, and more. We enumerate each of the threats presented in the catalog and add the following information:

1. Assignment of an attack vector to each threat, aligning with the CVSS schema discussed in Sect. 7.3.2.1.
2. Provision of an illustrative attack path as an example.
3. A description of the cybersecurity properties of an asset that may be compromised by the identified threat (Table A.4).

The evaluation presented below considers a single possible attack path and serves solely for illustrative purposes. The examples we provide are based on our experience and real-world cases. However, due to the general nature of the threat descriptions and the absence of a specific targeted asset, these examples may not universally apply to every item, component, asset, or TARA project. However, our goal is to provide guidance on how to evaluate these significant threats in the context of real-world projects, and to assist in their comprehensive analysis and understanding.

Table A.4 Threat catalog from UN Regulation No. 155, including our examples of attack path, cybersecurity properties, and attack vector

1 Back-end servers used as a means to attack a vehicle or extract data

Threat description (Vulnerability/Attack method)	1.1 Abuse of privileges by staff (insider attack)
Attack vector:	Network

Example of attack path [WP-15-05]:

1. The insider attack steals authentication credentials from a colleague to spoof the identity, or use two accounts in case the four-eyes principle login is activated

2. Log in to back-end system that manages the communication with the vehicles

3. The attacker sends valid requests to a vehicle (identified by a VIN) to activate paid features, or to retrieve sensitive data

Compromised cybersecurity property and reason [RQ-15-03]:

Confidentiality: extraction of vehicle-sensitive data by remote requests

Integrity: attack can be used to manipulate data

Availability: attack can be used to provoke denial of service (e.g., deactivate functions on vehicles)

Threat description (Vulnerability/Attack method)	1.2 Unauthorized Internet access to the server (enabled for example by backdoors, unpatched system software vulnerabilities, SQL attacks, or other means)
Attack vector:	Network

Example of attack path [WP-15-05]:

1. The attacker exploits a vulnerable back-end server to gain access to the server

2. Impersonation of the back-end server allows sending valid requests to vehicles

Compromised cybersecurity property and reason [RQ-15-03]:

Confidentiality: extraction of vehicle-sensitive data by remote requests

Integrity: attack can be used to manipulate data

Availability: attack can be used to provoke denial of service (e.g., deactivate functions on vehicles)

Threat description (Vulnerability/Attack method)	1.3 Unauthorized physical access to the server (conducted by for example USB sticks or other media connecting to the server)
Attack vector:	Physical

Example of attack path [WP-15-05]:

1. The attacker prepares malware on a bad USB (e.g., camouflaged as a keyboard) and manages to plug it into a critical back-end server

2. The malware gets introduced into the server

Compromised cybersecurity property and reason [RQ-15-03]:

Confidentiality: extraction of confidential data by getting full access to the server

Integrity: attack can be used to manipulate data

Availability: attack can be used to provoke a denial of service

(Continued)

Table A.4 (Continued)

2 Services from back-end server being disrupted, affecting the operation of a vehicle	
Threat description (Vulnerability/Attack method)	2.1 Attack on back-end server stops it functioning, for example it prevents it from interacting with vehicles and providing services they rely on
Attack vector:	Network
Example of attack path [WP-15-05]:	
1. The attacker has specific knowledge about the back-end services, such as protocols used, IP addresses, and basic protection mechanisms	
2. The attacker prepares a large-scale DDoS attack against the back-end	
Compromised cybersecurity property and reason [RQ-15-03]:	
Availability: the services provided by the back-end server are temporary disrupted	

3 Vehicle related data held on back-end servers being lost or compromised (data breach)	
Threat description (Vulnerability/Attack method)	3.1 Abuse of privileges by staff (insider attack)
Attack vector:	Local
Example of attack path [WP-15-05]:	
1. The insider attack steals authentication credentials from a colleague to spoof the identity, or use two accounts in case the four-eyes principle login is activated	
2. Log in to back-end system with enough privileges	
3. The attacker gets data from customers	
Compromised cybersecurity property and reason [RQ-15-03]:	
Confidentiality: extraction of sensitive customer's information	
Threat description (Vulnerability/Attack method)	3.2 Loss of information in the cloud. Sensitive data may be lost due to attacks or accidents when data is stored by third-party cloud service providers
Attack vector:	Network
Example of attack path [WP-15-05]:	
1. Attacker exploits a vulnerability and gets access to the cloud storage system	
2. The attacker executes malware or deletes files and backup data manually	
Compromised cybersecurity property and reason [RQ-15-03]:	
Integrity: the integrity of the data stored is directly compromised (data is deleted or modified) *Availability:* data loss	

(Continued)

Table A.4 (Continued)

3 Vehicle related data held on back-end servers being lost or compromised (data breach)	
Threat description (Vulnerability/Attack method)	3.3 Unauthorized Internet access to the server (enabled for example by backdoors, unpatched system software vulnerabilities, SQL attacks or other means)
Attack vector:	Network
Example of attack path [WP-15-05]:	
1. The attacker exploits a vulnerable back-end server	
2. Attacker can delete data on the server, or send fuzzed requests to corrupt the stored data	
Compromised cybersecurity property and reason [RQ-15-03]:	
Integrity: the vehicle data stored is deleted or modified	
Availability: data is not available due to deletion, or the data is corrupted	
Threat description (Vulnerability/Attack method)	3.4 Unauthorized physical access to the server (conducted for example by USB sticks or other media connecting to the server)
Attack vector:	Physical
Example of attack path [WP-15-05]:	
1. The attacker prepares malware over a bad USB (e.g., camouflaged as keyboard) and manages to plug it into a critical back-end server	
2. The malware gets introduced into the server	
Compromised cybersecurity property and reason [RQ-15-03]:	
Confidentiality: extraction of confidential data by full access to the server	
Integrity: attack can be used to manipulate data	
Availability: attack can be used to provoke denial of service	
Threat description (Vulnerability/Attack method)	3.5 Information breach by unintended sharing of data (e.g., admin errors)
Attack vector:	Network
Example of attack path [WP-15-05]:	
1. Due to bad policies and controls, an administrator sends sensitive data in plain-text format over the Internet	
Compromised cybersecurity property and reason [RQ-15-03]:	
Confidentiality: sensitive data is disclosed	

(Continued)

Table A.4 (Continued)

4 Spoofing of messages or data received by the vehicle	
Threat description (Vulnerability/ Attack method)	4.1 Spoofing of messages by impersonation (e.g., 802.11p V2X during platooning, GNSS messages, etc.)
Attack vector:	Adjacent
Example of attack path [WP-15-05]: 1. The attacker prepares jamming devices and spreads them over the city 2. The jamming devices have LTE connectivity and listen to a service operated by the attacker 3. The attacker launches a distributed jamming action to disrupt a particular service of the vehicle (e.g., rental company using remote service to unlock vehicles)	
Compromised cybersecurity property and reason [RQ-15-03]: *Integrity:* the messages between the vehicles and back-end servers are not reliable *Availability:* services relying on jammed signals get disrupted	
Threat description (Vulnerability/ Attack method)	4.2 Sybil attack (in order to spoof other vehicles as if there are many vehicles on the road)
Attack vector:	Adjacent
Example of attack path [WP-15-05]: 1. The attacker prepares transmission devices and spreads them over the city 2. The transmission devices impersonate vehicle presence by sending valid messages to other vehicles and infrastructure	
Compromised cybersecurity property and reason [RQ-15-03]: *Integrity:* V2X network is disrupted due to spoofed false identities *Availability:* services of V2X network might get temporary disrupted	

5 Communication channels used to conduct unauthorized manipulation, deletion or other amendments to vehicle held code/data	
Threat description (Vulnerability/ Attack method)	5.1 Communications channels permit code injection, for example tampered software binary might be injected into the communication stream
Attack vector:	Adjacent
Example of attack path [WP-15-05]: 1. The attacker connects to the ECU over an unprotected diagnostics or debugging interface 2. Binary of the ECU is extracted and reverse engineered 3. The attacker with enough capabilities (e.g., former employee or highly experienced SW engineer) builds a malware into the SW binary 4. The malware is introduced into the ECU	
Compromised cybersecurity property and reason [RQ-15-03]: *Confidentiality:* the malware can be used to exploit protections allowing to read out sensitive data *Integrity:* software of the ECU is modified *Availability:* the malware can make functions of the ECU unavailable *Authenticity:* the attacker can use the malware as a mean to bypass authentication methods (e.g., the ECU bypasses SecOC verification)	

(Continued)

Table A.4 (Continued)

5 Communication channels used to conduct unauthorized manipulation, deletion or other amendments to vehicle held code/data	
Threat description (Vulnerability/Attack method)	5.2 to 5.5 Communications channels permit manipulate/overwrite/erasure/write of vehicle held data/code
Attack vector:	Adjacent
Example of attack path [WP-15-05]:	
1. The attacker uses the OBD-II port as an attack vector	
2. The attacker is able to bypass diagnostics authentication mechanisms and gets elevated privileges to the target ECU	
3. Diagnostics messages are executed to read and write data	
Compromised cybersecurity property and reason [RQ-15-03]:	
Confidentiality: due to reading of data, in case that sensitive data is not protected enough	
Integrity: direct manipulation of data of the ECU	
Availability: the ECU can be completely disrupted	

6 Communication channels permit untrusted/unreliable messages to be accepted or are vulnerable to session hijacking/replay attacks	
Threat description (Vulnerability/Attack method)	6.1 Accepting information from an unreliable or untrusted source
Attack vector:	Adjacent
Example of attack path [WP-15-05]:	
1. Attacker connects to the ECU over an unprotected communication channel available	
2. Attacker sends valid requests to the ECU (e.g., crafting or replaying messages)	
3. The ECU accepts the requests without verification and answers by providing data	
Compromised cybersecurity property and reason [RQ-15-03]:	
Confidentiality: sensitive data can be disclosed by the requests	
Integrity: manipulated data can be introduced into the ECU	
Threat description (Vulnerability/Attack method)	6.2 Man in the middle attack/session hijacking
Attack vector:	Adjacent
Example of attack path [WP-15-05]:	
1. Attacker impersonates a cellular or Wi-Fi hotspot by using special equipment (evil twin) close to the vehicle	
2. The attacker sniffs all traffic transmitted from the vehicle, e.g., sensitive data such as username and passwords	
Compromised cybersecurity property and reason [RQ-15-03]:	
Confidentiality: possible disclosure of sensitive data	
Integrity: data in transit can be manipulated	
Authenticity: the attack can be used to steal authentication credentials and use them for a posterior attack	

(Continued)

Table A.4 (Continued)

6 Communication channels permit untrusted/unreliable messages to be accepted or are vulnerable to session hijacking/replay attacks	
Threat description (Vulnerability/Attack method)	6.3 Replay attack, for example an attack against a communication gateway allows the attacker to downgrade software of an ECU or firmware of the gateway
Attack vector:	Adjacent
Example of attack path [WP-15-05]: 1. The attacker compromises the communication unit of the vehicle 2. The malware in the communication unit is able to identify diagnostic messages used for software updates 3. When the gateway or target ECU requests an update, the malware at the communication unit modifies the request, announcing an update to an older software version	
Compromised cybersecurity property and reason [RQ-15-03]: *Integrity:* the integrity of the software update is compromised; the attacker uses known vulnerabilities at the downgraded software in order to execute further attacks	

7 Information can be readily disclosed. For example, through eavesdropping on communications or through allowing unauthorized access to sensitive files or folders	
Threat description (Vulnerability/Attack method)	7.1 Interception of information/interfering radiations/monitoring communications
Attack vector:	Adjacent
Example of attack path [WP-15-05]: 1. Attacker connects a sniffing device to the Ethernet network of the vehicle 2. Unprotected messages are captured by the sniffing device and stored	
Compromised cybersecurity property and reason [RQ-15-03]: *Confidentiality:* unprotected sensitive information is captured	
Threat description (Vulnerability/Attack method)	7.2 Gaining unauthorized access to files or data
Attack vector:	Adjacent
Example of attack path [WP-15-05]: 1. The attacker extracts an ECU from the vehicle and gains access to the board 2. The attacker identifies JTAG connectors on a chip and connects a debugging tool 3. Over the debugging tool, the attacker launches a brute-force attack to guess the JTAG password 4. The password is found, and the attacker gets access to the storage area on the chip	
Compromised cybersecurity property and reason [RQ-15-03]: *Confidentiality:* due to unauthorized access to data *Authenticity:* access protection broken	

(Continued)

Table A.4 (Continued)

8 Denial of service attacks via communication channels to disrupt vehicle functions	
Threat description (Vulnerability/Attack method)	8.1 Sending a large number of garbage data to vehicle information system, so that it is unable to provide services in the normal manner
Attack vector:	Adjacent
Example of attack path [WP-15-05]:	
1. The attacker uses a transmitting device close to a vehicle or group of vehicles (e.g., a device using stronger radiofrequency signals than the real Wi-Fi or V2X transmitters)	
2. The victim's vehicle receives false information and cannot operate normally (e.g., by spoofing GNSS information)	
Compromised cybersecurity property and reason [RQ-15-03]:	
Integrity: services relying on external communication become wrong information	
Availability: services relying on external communication are disrupted	
Threat description (Vulnerability/Attack method)	8.2 Black hole attack, in order to disrupt communication between vehicles the attacker is able to block messages between the vehicles
Attack vector:	Adjacent
Example of attack path [WP-15-05]:	
1. The attacker uses a transmitting device close to a vehicle or group of vehicles	
2. The attacker announces itself as a vehicle or communication node having the best route to any possible destination	
Compromised cybersecurity property and reason [RQ-15-03]:	
Integrity: wrong information is transmitted	
Availability: the communication relying on vehicle to vehicle communication is disrupted	

9 An unprivileged user is able to gain privileged access to vehicle systems	
Threat description (Vulnerability/Attack method)	9.1 An unprivileged user is able to gain privileged access, for example root access
Attack vector:	Adjacent
Example of attack path [WP-15-05]:	
1. The attacker extracts an ECU from the vehicle and gains access to the board	
2. The attacker identifies JTAG connectors on a board and connects a debugging tool	
3. Over the debugging tool, the attacker launches a brute-force attack to guess the JTAG password	
4. The password is found, and the attacker gets access to the storage area in the board	
Compromised cybersecurity property and reason [RQ-15-03]:	
Confidentiality: due to unauthorized access to data	
Authenticity: access protection broken	

(Continued)

Table A.4 (Continued)

10 Viruses embedded in communication media are able to infect vehicle systems	
Threat description (Vulnerability/Attack method)	10.1 Virus embedded in communication media infects vehicle systems
Attack vector:	Local
Example of attack path [WP-15-05]:	
1. An innocent user connects a USB device to the infotainment system	
2. The USB device contains a virus that is read by the infotainment system	
3. The virus presents itself as a software update, and it is executed	
Compromised cybersecurity property and reason [RQ-15-03]:	
Confidentiality: the malware can access to data stored in the infotainment system or at other ECUs by executing requests	
Integrity: software of the infotainment system is modified	
Availability: the malware can be used to disrupt certain services or functionalities	

11 Messages received by the vehicle (for example X2V or diagnostic messages), or transmitted within it, contain malicious content	
Threat description (Vulnerability/Attack method)	11.1 Malicious internal (e.g., CAN) messages
Attack vector:	Adjacent
Example of attack path [WP-15-05]:	
1. The attacker exploits a vulnerability in the Bluetooth stack of the vehicle	
2. The attacker can send commands that are accepted by the vehicle (e.g., leading to code execution)	
Compromised cybersecurity property and reason [RQ-15-03]:	
Confidentiality: in the worst case, unauthorized access to data is achieved	
Integrity: the attacker modifies the data of the vehicle	
Threat description (Vulnerability/Attack method)	11.2 Malicious V2X messages, e.g., infrastructure to vehicle or vehicle-vehicle messages (e.g., CAM, DENM)
Attack vector:	Adjacent
Example of attack path [WP-15-05]:	
1. The attacker uses a transmitting device close to a vehicle or group of vehicles (e.g., a device using stronger radiofrequency signals than the real Wi-Fi or V2X transmitters)	
2. The victim's vehicle receives false information and cannot operate normally (e.g., by spoofing GNSS information)	
Compromised cybersecurity property and reason [RQ-15-03]:	
Integrity: services relying on external communication become wrong information	
Availability: services relying on external communication are disrupted	

(Continued)

Table A.4 (Continued)

11 Messages received by the vehicle (for example X2V or diagnostic messages), or transmitted within it, contain malicious content	
Threat description (Vulnerability/Attack method)	11.3 Malicious diagnostic messages
Attack vector:	Adjacent
Example of attack path [WP-15-05]:	
1. The attacker exploits a vulnerable wireless interface (e.g., Bluetooth)	
2. Diagnostics messages are sent to the vehicle	
Compromised cybersecurity property and reason [RQ-15-03]:	
Confidentiality: malicious diagnostics messages are used to access data	
Integrity: malicious diagnostics messages can write data into the vehicle	
Threat description (Vulnerability/Attack method)	11.4 Malicious proprietary messages (e.g., those normally sent from OEM or component/system/function supplier)
Attack vector:	Adjacent
Example of attack path [WP-15-05]:	
1. The attacker exploits a vulnerable wireless interface (e.g., Bluetooth)	
2. Diagnostics messages are sent to the vehicle	
Compromised cybersecurity property and reason [RQ-15-03]:	
Confidentiality: malicious diagnostics messages are used to access data	
Integrity: malicious diagnostics messages can write data into the vehicle	

12 Misuse or compromise of update procedures	
Threat description (Vulnerability/Attack method)	12.1 Compromise of over the air software update procedures. This includes fabricating the system update program or firmware.
Attack vector:	Adjacent
Example of attack path [WP-15-05]:	
1. The attacker exploits a vulnerability in the head unit of a vehicle platform, allowing to install malware into the head unit	
2. The malware identifies software update checks from the ECUs of the vehicle	
3. When an ECU requests a software update, the malware on the head unit drops the request, or announces that there are no updates available	
Compromised cybersecurity property and reason [RQ-15-03]:	
Integrity: the whole update system is modified	
Availability: the update service is not available for the ECUs	

(Continued)

Table A.4 (Continued)

12 Misuse or compromise of update procedures	
Threat description (Vulnerability/Attack method)	12.2 Compromise of local/physical software update procedures. This includes fabricating the system update program or firmware
Attack vector:	Physical
Example of attack path [WP-15-05]: 1. The attacker has enough insider knowledge (e.g., a former employee) and can build an own software update image 2. The modified software update is introduced into the vehicle over the normal offline diagnostics tools	
Compromised cybersecurity property and reason [RQ-15-03]: *Integrity:* manipulated software is introduced into the vehicle	
Threat description (Vulnerability/Attack method)	12.3 The software is manipulated before the update process (and is therefore corrupted), although the update process is intact
Attack vector:	Local
Example of attack path [WP-15-05]: 1. The attacker installs a malware into the software update binary at the OEM or supplier back-end server 2. The modified software update has valid signatures, and it is installed into the vehicles using standard over-the-air update mechanisms	
Compromised cybersecurity property and reason [RQ-15-03]: *Confidentiality:* the manipulated code can get unauthorized access to data in the vehicle *Integrity:* manipulated software is introduced into the vehicle	
Threat description (Vulnerability/Attack method)	12.4 Compromise of cryptographic keys of the software provider to allow invalid update
Attack vector:	Adjacent
Example of attack path [WP-15-05]: 1. An internal employee, or an attacker getting enough access to the software build environment, uses the signing process to sign an own software binary 2. The attacker uses the valid software binary for introducing it into the vehicle	
Compromised cybersecurity property and reason [RQ-15-03]: *Confidentiality:* unauthorized access to data possible (e.g., software might get access to protected areas) *Integrity:* manipulated software is introduced into the vehicle	

(Continued)

Table A.4 (Continued)

13 It is possible to deny legitimate updates	
Threat description (Vulnerability/Attack method)	13.1 Denial of Service attack against update server or network to prevent rollout of critical software updates and/or unlock of customer specific features
Attack vector:	Network
Example of attack path [WP-15-05]:	
1. The attacker has specific knowledge about the back-end services, such as protocols used, IP addresses, and basic protection mechanisms	
2. The attacker prepares a large-scale DDoS attack against the back-end	
Compromised cybersecurity property and reason [RQ-15-03]:	
Availability: the software update services are temporary disrupted	
15[a] Legitimate actors are able to take actions that would unwittingly facilitate a cyber-attack	
Threat description (Vulnerability/Attack method)	15.1 Innocent victim (e.g., owner, operator or maintenance engineer) being tricked into taking an action to unintentionally load malware or enable an attack
Attack vector:	Local
Example of attack path [WP-15-05]:	
1. The attacker compromises the customer service of an OEM and retrieves a list of customers, including e-mail addresses	
2. The attacker sends a phishing attack to each customer, including real names and VINs	
3. The phishing mail explains that to use some new features of the vehicle, the user has to download an update provided by a link and download it to the car over USB	
4. The user downloads the malware from the link and inserts it into the vehicle	
Compromised cybersecurity property and reason [RQ-15-03]:	
Integrity: malicious software is introduced into the vehicle	
Threat description (Vulnerability/Attack method)	15.2 Defined security procedures are not followed
Attack vector:	Local
Example of attack path [WP-15-05]:	
1. The procedure for assigning series keys in production through a configuration in the BOM has not been executed correctly	
2. The ECU is produced with series keys and keys used only for development	
3. The attacker, who is aware of the vulnerability, can access to the ECU by using development keys	
Compromised cybersecurity property and reason [RQ-15-03]:	
Confidentiality: unauthorized access to data possible	
Integrity: data/software modification	
Authenticity: authentication mechanisms are bypassed, or not used as expected	

(Continued)

Table A.4 (Continued)

16 Manipulation of the connectivity of vehicle functions enables a cyber-attack, this can include telematics; systems that permit remote operations; and systems using short range wireless communications	
Threat description (Vulnerability/ Attack method)	16.1 Manipulation of functions designed to remotely operate systems, such as remote key, immobilizer, and charging pile
Attack vector:	Local
Example of attack path [WP-15-05]: 1. The attacker performs SW manipulation of the connectivity unit of the vehicle 2. The installed malware is able to control remote functions, e.g., activate or deactivate functions when the attacker wants	
Compromised cybersecurity property and reason [RQ-15-03]: *Integrity:* software is manipulated *Availability:* remote functions partially or complete not available	
Threat description (Vulnerability/ Attack method)	16.2 Manipulation of vehicle telematics (e.g., manipulate temperature measurement of sensitive goods, remotely unlock cargo doors)
Attack vector:	Local
Example of attack path [WP-15-05]: 1. The attacker performs SW manipulation of the telematics unit 2. A malware installed modifies the messages received by the telematics unit	
Compromised cybersecurity property and reason [RQ-15-03]: *Integrity:* software is manipulated	
Threat description (Vulnerability/ Attack method)	16.3 Interference with short range wireless systems or sensors
Attack vector:	Adjacent
Example of attack path [WP-15-05]: 1. The attacker performs MITM on short-range wireless systems and compromises unprotected messages 2. The attacker can craft messages impersonating sensors, and the vehicle receives the wrong sensors measurements	
Compromised cybersecurity property and reason [RQ-15-03]: *Integrity:* sensors or telematics messages are manipulated	

(Continued)

Table A.4 (Continued)

17 Hosted 3rd party software, e.g., entertainment applications, used as a means to attack vehicle systems

Threat description (Vulnerability/Attack method)	17.1 Corrupted applications, or those with poor software security, used as a method to attack vehicle systems
Attack vector:	Local

Example of attack path [WP-15-05]:

1. The attacker creates a malicious application and makes it available over the application distribution platform used in the vehicle

2. The end user installs the application and uses it normally

3. The malicious application runs a small service that allows the attacker to get access to the vehicle

Compromised cybersecurity property and reason [RQ-15-03]:

Confidentiality: the attacker might get access to sensitive data stored in the vehicle

Integrity: the malicious application or the attacker can compromise other applications or the complete vehicle

Availability: some functions of the vehicle might get disturbed due to a follow-up attack enabled by the malware

18 Devices connected to external interfaces, e.g., USB ports, OBD port, used as a means to attack vehicle systems

Threat description (Vulnerability/Attack method)	18.1 External interfaces such as USB or other ports used as a point of attack, for example through code injection
Attack vector:	Physical

Example of attack path [WP-15-05]:

1. A cheap USB or OBD dongle used for diagnostics purposes contains malware and connectivity capabilities

2. The end user plugs in the dongle, and malware is automatically installed in other ECUs of the vehicle

3. The dongle, with its connectivity capabilities and the installed malware, provide the attacker with remote access to the vehicle

Compromised cybersecurity property and reason [RQ-15-03]:

Confidentiality: the attacker might get access to sensitive data stored in the vehicle

Integrity: the malicious application or the attacker can compromise other applications or the complete vehicle

Availability: some functions of the vehicle might get disturbed due to a follow attack-up enabled by the malware

(Continued)

Table A.4 (Continued)

18 Devices connected to external interfaces, e.g., USB ports, OBD port, used as a means to attack vehicle systems	
Threat description (Vulnerability/Attack method)	18.2 Media infected with a virus connected to a vehicle system
Attack vector:	Physical
Example of attack path [WP-15-05]: 1. The end user installs an application, or inserts a media device, containing malware 2. The malicious application runs a small service that allows the attacker to get access to the vehicle	
Compromised cybersecurity property and reason [RQ-15-03]: *Confidentiality:* the attacker might get access to sensitive data stored in the vehicle *Integrity:* the malicious application or the attacker can compromise other applications or the complete vehicle *Availability:* some functions of the vehicle might get disturbed due to a follow-up attack enabled by the malware	
Threat description (Vulnerability/Attack method)	18.3 Diagnostic access (e.g., dongles in OBD port) used to facilitate an attack, e.g., manipulate vehicle parameters (directly or indirectly)
Attack vector:	Local
Example of attack path [WP-15-05]: 1. A cheap OBD dongle used for diagnostics purposes contains malware and connectivity capabilities 2. The end user plugs in the dongle, and malware is automatically installed in other ECUs of the vehicle 3. The dongle, with its connectivity capabilities and the installed malware, provide the attacker with remote access to the vehicle	
Compromised cybersecurity property and reason [RQ-15-03]: *Confidentiality:* the attacker might get access to sensitive data stored in the vehicle *Integrity:* the malicious application or the attacker can compromise other applications or the complete vehicle *Availability:* some functions of the vehicle might get disturbed due to a follow-up attack enabled by the malware	

19 Extraction of vehicle data/code	
Threat description (Vulnerability/Attack method)	19.1 Extraction of copyright or proprietary software from vehicle systems (product piracy)
Attack vector:	Physical
Example of attack path [WP-15-05]: 1. The attacker unmounts an ECU from the vehicle 2. The attacker connects to the ECU and gets access to unprotected data and software binary 3. Reverse engineering is performed on the binary	
Compromised cybersecurity property and reason [RQ-15-03]: *Confidentiality:* The working principles of the software are revealed to the attacker. Additionally, sensitive data might get read.	

(Continued)

Table A.4 (Continued)

19 Extraction of vehicle data/code	
Threat description (Vulnerability/Attack method)	19.2 Unauthorized access to the owner's privacy information such as personal identity, payment account information, address book information, location information, vehicle's electronic ID, etc.
Attack vector:	Physical
Example of attack path [WP-15-05]:	
1. The attacker unmounts an ECU from the vehicle	
2. The attacker connects to the ECU and gets access to unprotected data	
Compromised cybersecurity property and reason [RQ-15-03]:	
Confidentiality: due to unauthorized access to PII, or other sensitive data	
Threat description (Vulnerability/Attack method)	19.3 Extraction of cryptographic keys
Attack vector:	Physical
Example of attack path [WP-15-05]:	
1. Local access to the ECU	
2. Diagnostics or debugging interface/protocol is exploited	
3. Not well protected cryptographic keys are extracted (e.g., credentials for communication to back-end servers)	
Compromised cybersecurity property and reason [RQ-15-03]:	
Confidentiality: unauthorized access to sensitive information	
Authenticity: due to unauthorized access over an exploited interface	

20 Manipulation of vehicle data/code	
Threat description (Vulnerability/Attack method)	20.1 Illegal/unauthorized changes to vehicle's electronic ID
Attack vector:	Physical
Example of attack path [WP-15-05]:	
1. The attacker gets unauthorized access to the unprotected diagnostics interface of some ECU	
2. Due to the elevation of privileges, the attacker can perform modifications on the data stored	
Compromised cybersecurity property and reason [RQ-15-03]:	
Integrity: modification of data	
Availability: genuine ID gets modified and is not available anymore	
Authenticity: due to unauthorized access over not enough protected interface	

(Continued)

Table A.4 (Continued)

20 Manipulation of vehicle data/code	
Threat description (Vulnerability/Attack method)	20.2 Identity fraud. For example, if a user wants to display another identity when communicating with toll systems, manufacturer backend
Attack vector:	Physical
Example of attack path [WP-15-05]: 1. The attacker gets unauthorized access to the unprotected diagnostics interface of some ECU 2. Due to the elevation of privileges, the attacker can perform modifications on the data stored 3. The attacker impersonates another valid identity and communicates with the back-end systems	
Compromised cybersecurity property and reason [RQ-15-03]: *Integrity:* modification of data *Authenticity:* due to unauthorized access over a not-enough-protected interface	
Threat description (Vulnerability/Attack method)	20.3 Action to circumvent monitoring systems (e.g., hacking/tampering/blocking of messages such as ODR Tracker data, or number of runs)
Attack vector:	Physical
Example of attack path [WP-15-05]: 1. The attacker gets unauthorized access to the unprotected diagnostics interface of some ECU 2. Due to the elevation of privilegesn the attacker can perform modifications on the data stored	
Compromised cybersecurity property and reason [RQ-15-03]: *Integrity:* modification of data *Availability:* genuine ID gets modified and is not available anymore *Authenticity:* due to unauthorized access over a not-enough-protected interface	
Threat description (Vulnerability/Attack method)	20.4 Data manipulation to falsify vehicle's driving data (e.g., mileage, driving speed, driving directions, etc.)
Attack vector:	Physical
Example of attack path [WP-15-05]: 1. The attacker installs a malware over the infotainment ECU (e.g., over a USB media) 2. The malware infects other ECUs and becomes write-access 3. Additionally, the malware is able to inject manipulated messages on the vehicle bus	
Compromised cybersecurity property and reason [RQ-15-03]: *Integrity:* software and data on the vehicle are modified. Moreover, wrong driving data is provided. *Availability:* genuine driving data is not available	
Threat description (Vulnerability/Attack method)	20.5 Unauthorized changes to system diagnostic data
Attack vector:	Physical
Example of attack path [WP-15-05]: 1. The attacker gets unauthorized access to the unprotected diagnostics interface of some ECU 2. Due to the elevation of privileges, the attacker can perform modifications on the data stored	
Compromised cybersecurity property and reason [RQ-15-03]: *Integrity:* modification of data *Authenticity:* due to unauthorized access over a not-enough-protected interface	

(Continued)

Table A.4 (Continued)

21 Erasure of data/code

Threat description (Vulnerability/Attack method)	21.1 Unauthorized deletion/manipulation of system event logs
Attack vector:	Physical

Example of attack path [WP-15-05]:

1. The attacker gets unauthorized write access to the unprotected diagnostics interface of some ECU
2. The attacker modifies event logs of the ECU or several ECUs

Compromised cybersecurity property and reason [RQ-15-03]:

Integrity: modification of data

Availability: due to deletion of event logs

Authenticity: due to unauthorized access over a not-enough-protected interface

22 Introduction of malware

Threat description (Vulnerability/Attack method)	22.2 Introduce malicious software or malicious software activity
Attack vector:	Physical

Example of attack path [WP-15-05]:

1. The attacker connects to the ECU over an unprotected communication channel available
2. The attacker modifies the SW binary or introduces malware

Compromised cybersecurity property and reason [RQ-15-03]:

Integrity: manipulation of software

Availability: genuine SW binary is not available

Authenticity: due to unauthorized access over not enough protected interface

23 Introduction of new software or overwrite existing software

Threat description (Vulnerability/Attack method)	23.1 Fabrication of software of the vehicle
Attack vector:	Physical

Example of attack path [WP-15-05]:

1. The attacker connects to ECU over an unprotected communication channel available
2. The attacker creates valid SW binary
3. SW binary is injected into the ECU

Compromised cybersecurity property and reason [RQ-15-03]:

Integrity: manipulation of software

Availability: genuine SW binary not available

Authenticity: due to unauthorized access over not enough protected interface

(Continued)

Table A.4 (Continued)

24 Disruption of systems or operations

Threat description (Vulnerability/Attack method)	24.1 Denial of service, for example this may be triggered on the internal network by flooding a CAN bus, or by provoking faults on an ECU via a high rate of messaging
Attack vector:	Adjacent

Example of attack path [WP-15-05]:

1. The attacker gets access to the vehicle CAN bus
2. Over a manipulated ECU or a special device, the attacker performs flooding on the CAN bus

Compromised cybersecurity property and reason [RQ-15-03]:

Availability: in-vehicle communication not possible due to flooded bus

25 Manipulation of vehicle parameters

Threat description (Vulnerability/Attack method)	25.1 Unauthorized access of falsify the configuration parameters of vehicle's key functions, such as brake data, airbag deployed threshold, etc.
Attack vector:	Physical

Example of attack path [WP-15-05]:

1. The attacker installs a malware over the infotainment ECU (e.g., over a USB media)
2. The malware modifies vehicle's function at some ECUs

Integrity: software and data on the vehicle is modified

Availability: genuine data is not available

Threat description (Vulnerability/Attack method)	25.2 Unauthorized access of falsify the charging parameters, such as charging voltage, charging power, battery temperature, etc.
Attack vector:	Physical

Example of attack path [WP-15-05]:

1. The attacker connects to ECU over an unprotected communication channel available
2. The attacker can modify data at the ECU

Compromised cybersecurity property and reason [RQ-15-03]:

Integrity: due to modification of data

Availability: genuine data is not available

Authenticity: due to unauthorized access over not enough protected interface

(Continued)

Table A.4 (Continued)

26 Cryptographic technologies can be compromised or are insufficiently applied	
Threat description (Vulnerability/Attack method)	26.1 Combination of short encryption keys and long period of validity enables attacker to break encryption
Attack vector:	Physical
Example of attack path [WP-15-05]:	
1. The attacker exploits cryptography keys and algorithms	
2. The attacker gets full access to the device, e.g., for writing new software	
Compromised cybersecurity property and reason [RQ-15-03]:	
Confidentiality: unauthorized access to sensitive data due to broken encryption	
Integrity: the attacker can modify data in a subsequent action	
Availability: ECU or data on the ECU can get unavailable	
Authenticity: unauthorized access due to broken cryptography	
Threat description (Vulnerability/Attack method)	26.2 Insufficient use of cryptographic algorithms to protect sensitive systems
Attack vector:	Physical
Example of attack path [WP-15-05]:	
1. The attacker exploits cryptography keys and algorithms	
2. The attacker gets full access to the device, e.g., for writing new software	
Compromised cybersecurity property and reason [RQ-15-03]:	
Confidentiality: unauthorized access to sensitive data due to broken encryption	
Integrity: the attacker can modify data in a subsequent action	
Availability: ECU or data on the ECU can get unavailable	
Authenticity: unauthorized access due to broken cryptography	
Threat description (Vulnerability/Attack method)	26.3 Using already or soon to be deprecated cryptographic algorithms
Attack vector:	Physical
Example of attack path [WP-15-05]:	
1. The attacker exploits cryptography keys and algorithms	
2. The attacker gets full access to the device, e.g., for writing new software	
Compromised cybersecurity property and reason [RQ-15-03]:	
Confidentiality: unauthorized access to sensitive data due to broken encryption	
Integrity: the attacker can modify data in a subsequent action	
Availability: ECU or data on the ECU can get unavailable	
Authenticity: unauthorized access due to broken cryptography	

(Continued)

Table A.4 (Continued)

27 Parts or supplies could be compromised to permit vehicles to be attacked	
Threat description (Vulnerability/Attack method)	27.1 Hardware or software, engineered to enable an attack or fails to meet design criteria to stop an attack
Attack vector:	Physical
Example of attack path [WP-15-05]:	
1. The attacker gets insider information about technical operational details of a vehicle function	
2. The attacker writes a malware that is able to bypass insufficient protection countermeasures	
3. The vehicle function gets compromised	
Compromised cybersecurity property and reason [RQ-15-03]:	
Integrity: software or data (e.g., parameters) are modified	
Availability: vehicle gets partially or completely out of service	

28 Software or hardware development permits vulnerabilities	
Threat description (Vulnerability/Attack method)	28.1 Software bugs. The presence of software bugs can be a basis for potential exploitable vulnerabilities. This is particularly true if software has not been tested to verify that known bad code/bugs is not present and reduce the risk of unknown bad code/bugs being present
Attack vector:	Physical
Example of attack path [WP-15-05]:	
1. The attacker performs intensive research on a particular ECU until a bug is found	
2. The attacker writes a malware for exploiting the ECU and gets writing access	
3. A modified SW binary is introduced into the ECU, compromising the rest of the vehicle	
Compromised cybersecurity property and reason [RQ-15-03]:	
Confidentiality: sensitive data might get disclosed	
Integrity: SW is manipulated or malware is introduced	
Availability: vehicle gets partially or completely out of service	
Authenticity: authentication mechanisms get exploited or bypassed	
Threat description (Vulnerability/Attack method)	28.2 Using remainders from development (e.g., debug ports, JTAG ports, microprocessors, development certificates, developer passwords, …) can permit access to ECUs or permit attackers to gain higher privileges
Attack vector:	Physical
Example of attack path [WP-15-05]:	
1. The attacker performs intensive research on a particular ECU until a bug is found	
2. The attacker can exploit a debugging interface and gets writing access to the ECU	
3. A modified SW binary is introduced into the ECU, compromising the rest of the vehicle	
Compromised cybersecurity property and reason [RQ-15-03]:	
Confidentiality: sensitive data might get disclosed	
Integrity: SW is manipulated or malware is introduced	
Availability: vehicle gets partially or completely out of service	
Authenticity: authentication mechanisms get exploited or bypassed	

(Continued)

Table A.4 (Continued)

29 Network design introduces vulnerabilities	
Threat description (Vulnerability/Attack method)	29.1 Superfluous Internet ports left open, providing access to network systems
Attack vector:	Network
Example of attack path [WP-15-05]:	
1. The attacker performs a reconnaissance and port scan on the vehicle's head unit	
2. A powerful port, such as for SSH, is open, and the attacker connects to it	
Compromised cybersecurity property and reason [RQ-15-03]:	
Confidentiality: sensitive data might get disclosed	
Integrity: SW is manipulated or malware is introduced	
Availability: vehicle gets partially or completely out of service	
Threat description (Vulnerability/Attack method)	29.2 Circumvent network separation to gain control. Specific example is the use of unprotected gateways, or access points (such as truck-trailer gateways), to circumvent protections and gain access to other network segments to perform malicious acts, such as sending arbitrary CAN bus messages
Attack vector:	Adjacent
Example of attack path [WP-15-05]:	
1. The attacker connects an external device to the vehicle network (e.g., CAN or Ethernet network)	
2. A vulnerability or wrong configuration at the vehicle gateway allows getting access to ECUs behind the gateway	
3. The attacker can insert code or send unauthorized signals to the ECUs	
Compromised cybersecurity property and reason [RQ-15-03]:	
Integrity: vehicle or ECUs can get damaged	
Availability: vehicle gets partially or completely out of service	

30 Threats of physical loss of data	
Threat description (Vulnerability/Attack method)	30.1 Damage caused by a third party. Sensitive data may be lost or compromised due to physical damages in cases of traffic accident or theft
Attack vector:	Physical
Example of attack path [WP-15-05]:	
1. The attacker physically connects to memory are and reads out sensitive data	
Compromised cybersecurity property and reason [RQ-15-03]:	
Confidentiality: unauthorized access to sensitive data	

(Continued)

Table A.4 (Continued)

30 Threats of physical loss of data	
Threat description (Vulnerability/Attack method)	30.2 Loss from DRM (digital right management) conflicts. User data may be deleted due to DRM issues
Attack vector:	Physical
Example of attack path [WP-15-05]:	
1. The attacker detaches an ECU such as the infotainment	
2. The attacker possesses confidential information about the HW of the ECU	
3. The attacker makes a HW modification to the memory area where the DRM information is stored	
Compromised cybersecurity property and reason [RQ-15-03]:	
Integrity: due to physical modification	
Availability: DRM information not available any more	
Threat description (Vulnerability/Attack method)	30.3 The (integrity of) sensitive data may be lost due to IT components wear and tear, causing potential cascading issues (in case of key alteration, for example)
Attack vector:	Physical
Example of attack path [WP-15-05]:	
1. The attacker is an insider attack that knows the problematic of IT components not working properly	
2. The insider attacker takes advantage of it and is able to create valid tokens for activating features in vehicles	
3. The tokens are sold on illegal market	
Compromised cybersecurity property and reason [RQ-15-03]:	
Integrity: feature activation mechanism gets corrupted	
Authenticity: authentication mechanisms on older devices get bypassed	

31 Unintended transfer of data can occur	
Threat description (Vulnerability/Attack method)	31.1 Information breach. Personal data may be leaked when the car changes user (e.g., is sold or is used as hire vehicle with new hirers)
Attack vector:	Physical
Example of attack path [WP-15-05]:	
1. The attacker performs an intensive penetration test on the vehicle	
2. Personal data of older users is still stored at some ECUs for different purposes	
3. The attacker finally gets access to the PII	
Compromised cybersecurity property and reason [RQ-15-03]:	
Confidentiality: unauthorized access to PII	
Authenticity: broken or insufficient authentication mechanisms	

(Continued)

Table A.4 (Continued)

32 Physical manipulation of systems can enable an attack	
Threat description (Vulnerability/Attack method)	32.1 Manipulation of electronic hardware, e.g., unauthorized electronic hardware added to a vehicle to enable man-in-the-middle attack; Replacement of authorized electronic hardware (e.g., sensors) with unauthorized electronic hardware; Manipulation of the information collected by a sensor (for example, using a magnet to tamper with the Hall effect sensor connected to the gearbox)
Attack vector:	Physical
Example of attack path [WP-15-05]:	
1. The attacker opens the ECU of a target vehicle	
2. A manipulated SW is installed into the ECU, or an additional electronic device is mounted so that it can inject or modify data in transit	
Compromised cybersecurity property and reason [RQ-15-03]:	
Integrity: vehicle or ECUs can get damaged	
Availability: vehicle gets partially or completely out of service	

[a]The attentive reader may wonder why the number 14 is missing. This is intentional, as the threat number 14 is indeed not included in the document of UN Regulation No. 155.

Index

R. do Carmo and A. Schlensog, *Automotive Threat Analysis and Risk Assessment in Practice*, https://doi.org/10.1007/978-3-662-69614-9

The manufacturer's authorised representative in the EU is Springer Nature Customer Service Centre GmbH, Europaplatz 3, 69115 Heidelberg, Germany. If you have any concerns regarding our products, please contact ProductSafety@springernature.com

Printed and bound by CPI Group (UK) Ltd, Croydon, CR0 4YY
30/09/2026
02225286-0018